Plastics for Schools

PLASTICS FOR SCHOOLS

APPLIED POLYMER SCIENCE

PETER J CLARKE

MILLS & BOON LIMITED London

First published 1970 by Allman & Son Ltd
(now incorporated in Mills & Boon Ltd)
17–19 Foley Street, London W1A 1DR
2nd edition 1972
3rd edition 1976

© Peter J. Clarke 1970

ISBN 0 263 05380 6

Printed in Great Britain by
Butler & Tanner Ltd, Frome and London

Contents

List of Plates

List of Diagrams

ix

Acknowledgements

I am greatly indebted to Hillingdon Education Authority for the help they have shown in the preparation of this book, and in particular to Mr P. J. Care, Adviser for Technical Subjects and Craft, for his long-standing assistance and encouragement in this field. I would also like to acknowledge the help of several companies in the Plastics Industry, whose very generous co-operation in the supply of technical assistance and materials for experimental work, as well as photographs for the book, has been so valuable. I am grateful to Mr H. S. Finlay, B.Sc., for contributing the first section on the Chemistry of Plastics, to Mr Bob Beddis for his very willing help in taking photographs and to Mr David Smith for his help and advice during the preparation of the manuscript.

Publishers' Note

The Publishers feel that for the present a mixture of traditional and metric units will be acceptable. It is desirable that in the various measurements relating to school-made equipment and calculations of temperature metric units should be employed.

On the other hand, where it seems that conversion is likely to impair the understanding of various examples or comparisons, also where some descriptions of existing machinery are involved, traditional units are probably a more realistic mode of expression.

Foreword

There is an urgent need to bring an awareness of all classes
of materials into schools' curricula. For many years projects
have been devised in which articles, machines, etc., are
made up from metal, wood and other materials, but since
polymers are becoming so important in our lives it is essential
for people at school to know their capabilities and uses. This
is particularly important since plastics were introduced as
large-scale tonnage materials in the role of substitutes for
'better' traditional materials which were either too expensive
or unobtainable. This bad image has been perpetuated and
even increased by a tendency to consider only the piecemeal
replacement of components by ones made from plastics.
New materials need new designs—especially since there are
an increasing number of applications where plastics are not
only the best but indeed the only suitable materials.

Mr Clarke's book does three things towards encouraging this
attitude of mind. Firstly, it helps one to understand how
plastics and rubbers are made up and how different properties
can be designed into them. Secondly, it demonstrates the
effect of various conditions on the properties of plastics,
showing how they will behave in different environments.
Finally, it gives guidance on the most appropriate use of
different types of plastics in carrying out specific projects.

W. A. Holmes-Walker
Director-General,
British Plastics Federation,
formerly Professor and Head of Department of Polymer
Science and Technology, Brunel University

Introduction

Plastics are essentially materials of the twentieth century although their development can be traced back for just a hundred years. A tremendous growth has been evident in the last thirty years and now plastics rank second only to metals in their usefulness as engineering materials.

Plastics materials possess a wide variation in their properties and this is what gives them their versatility and adaptability. Some are soft and flexible, whereas others are hard and can be worked like metal or wood; others combine hardness with resilience and some hardness with brittleness. Most can be moulded into quite complicated shapes. Generally, plastics are not as strong as metals but being less dense their strength-to-weight ratio is quite good; strength can be added by the inclusion of glass and now carbon fibres in the plastics material. Most plastics are good thermal and electrical insulators and possess resistance to corrosion and to chemical attack. Some plastics have very special properties such as a very low coefficient of friction, or the ability to flex back and forth many thousands of times, or excellent light-transmission properties. Colours can be incorporated into plastics and many can also be produced in transparent form.

They should not be thought of as one material but as a very diverse family of thirty or more in number. Some of these derive from natural materials such as cellulose, casein and rubber, whereas others are totally man-made by a building-up process from simple chemical substances.

Today plastics find applications in much of our day-to-day life: the light switch, the tooth-brush, the door handle, the ball-point pen, our clothing, food packaging and space travel all demonstrate plastics materials of many different types in use.

It is desirable, therefore, that boys and girls of all abilities should have some experience of these materials together with the conventional school craft materials—metal, wood, clay, fabrics and paint. It is hoped that this book will provide an opportunity for teachers and pupils to work with plastics and the associated processes, and to discover the scope and the limitations of the materials.

What plastics can do and how they behave in use depends on their chemical structure, and this is discussed together with basic techniques in the production and processing of plastics. There follow three sections containing some five topics each, taking the reader through a number of simple activities using plastics materials to the use of small industrial-type equipment.

The family of plastics materials as a whole can help promote many ideas for practical work in schools, broadening its scope immensely. It is not intended to convey the idea that plastics are here to replace all other materials, but they should be available if the design and fabrication problems of the art, craft, technical or any of the other departments of a school require their use. Plastics will be seen for what they are—a fascinating and highly useful group of materials.

Section I

THE CHEMISTRY OF PLASTICS

The Chemistry of Plastics

The term 'plastics' covers a large group of materials which exhibit a great diversity in their physical and chemical properties. It is this diversity in properties which makes them so useful, and manufacturing processes have been so regulated by factors such as temperature, pressure and catalysts that plastics materials can be produced to meet specific requirements. It is not surprising therefore that the traditional materials, metal and wood, have been replaced to such a large extent by plastics.

It is anticipated that many users of this book will have a limited knowledge of chemistry and therefore a limited capacity for understanding the relationship between the properties of plastics and their structure. The purpose of this chapter is to present, in as simple a manner as possible, a chemical treatment which, it is hoped, will promote a fuller understanding.

1 Molecules and Atoms

All matter is composed of simple substances known as *elements*. An *element* may be defined as a substance which cannot be split up chemically into simpler substances. All the metals are elements. However, the elements most commonly found in plastics materials are carbon, hydrogen, oxygen, nitrogen, fluorine and chlorine. The smallest particle of an element which can take part in a chemical reaction is known as an *atom*. Atoms are represented by appropriate symbols, e.g.

C, H, O, N represent respectively atoms of carbon, hydrogen, oxygen and nitrogen.

When two or more different elements combine chemically a
compound is formed. Water is a compound which is produced
when hydrogen and oxygen react. Every oxygen atom is united
by chemical bonds to two hydrogen atoms, producing what is
known as a **molecule** of water, which is represented by the
formula H_2O. It is a comparatively simple operation to split
water again into its constituent elements, hydrogen and
oxygen, but it is not possible to split up the hydrogen and
the oxygen into any simpler substances. A **molecule** is the
smallest particle of an element or a compound that can
lead a normal, separate existence.

It is often more useful to represent a substance by its structural
formula than by its molecular formula, e.g., in the case of
water,

H_2O

Molecular formula   Structural formula

The following points concerning this structural formula must
be made clear at this stage:

(a) The water molecule is planar and is V-shaped.
(b) The strokes between atoms represent chemical bonds.
(c) Since one oxygen atom can combine with two hydrogen atoms
it is said to have a **valency** of two, i.e. it is divalent. The
hydrogen atom is monovalent. Since the type of chemical
bond occurring in plastics materials is usually the same as that
shown for water, it is reasonably safe to assume that the
valency of an atom is numerically equal to the number of
bonds (strokes) emanating from the symbol for that atom in
the structural formula.
(d) The bond angle HOH is 104° 31′. From this it will be seen
that this type of chemical bond is strictly directional.

It has to be remembered that the structures of compounds are
usually **three**-dimensional arrays of the atoms of their
constituent elements so that obvious limitations are imposed

3

by a two-dimensional graphical representation. The following examples should clarify this:

H—N (with H above and H below) — Ammonia (usual structural representation)

H----N (with H and H below) (HNH = 107°) — Ammonia (actual structure)

The significance of the broken line in the actual structure is that the chemical bond is directed below the plane of the paper.

The symbol ——◀ represents a bond directed above the plane of the paper.

A stroke of uniform width (——) represents a bond in the plane of the paper.

H—C—H (with H above and H below) — Methane (structural formula)

H—C----H (with H above and H below) (HCH = 109° 28′) — Methane (actual structure)

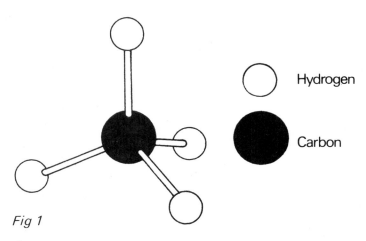

○ Hydrogen

● Carbon

Fig 1

It would be advisable to obtain a set of atomic models of the ball-and-spoke type from the Chemistry Department and to construct, from them, molecular models of water, ammonia and methane. It is worth noting that in the last two mentioned compounds the valencies of the nitrogen and carbon atoms are three and four respectively. A model of the methane molecule is shown in fig 1.

Methane is the simplest member of a family of chemically related hydrocarbons known as **alkanes**. The general formula for the paraffins is C_nH_{2n+2}. The molecular and structural formulae of the simpler paraffins are given below:

Alkanes	Molecular formula	Structural formula
Methane	CH_4	
Ethane	C_2H_6	
Propane	C_3H_8	
Butane	C_4H_{10}	
2-methyl propane	C_4H_{10}	

Methane:
```
        H
        |
   H —— C —— H
        |
        H
```

Ethane:
```
     H   H
     |   |
 H — C — C — H
     |   |
     H   H
```

Propane:
```
     H   H   H
     |   |   |
 H — C — C — C — H
     |   |   |
     H   H   H
```

Butane:
```
     H   H   H   H
     |   |   |   |
 H — C — C — C — C — H
     |   |   |   |
     H   H   H   H
```

2-methyl propane:
```
     H           H           H
     |           |           |
 H — C ————————— C ————————— C — H
     |           |           |
     H       H — C — H       H
                 |
                 H
```

5

The chains of carbon atoms in ethane, propane and butane appear straight when represented structurally. Molecular models of these compounds should be constructed and examined. This will reveal (a) that the carbon chains are not straight and (b) that rotation is possible about a **single** bond between adjacent carbon atoms. Isobutane differs from butane in that it is a branched chain compound while butane is a so-called 'straight chain' compound. Compounds, such as these, which have the same molecular formula but different structural formulae are known as **isomers** and the phenomenon is known as **isomerism**.

2 Saturation and Unsaturation

Organic (carbon) compounds, such as the paraffins, in which each carbon atom in the molecule is linked by **single** bonds to four other atoms are known as saturated compounds. If, however, an organic compound has in its molecule one or more carbon atoms linked chemically by double (or triple) bonds to some other atom then that compound is said to be unsaturated. Ethylene is a typical unsaturated compound having the following structural formula:

$$\underset{H}{\overset{H}{\diagdown}} C = C \underset{H}{\overset{H}{\diagup}}$$

Examination of a molecular model reveals that (a) the ethylene molecule is planar and (b) that there is no rotation about a double bond. One of the bonds between the carbon atoms is weaker than the other, so that under suitable experimental

6

conditions the weaker bond is ruptured, creating conditions under which an *addition reaction* may take place:

Ethylene   Chlorine   1,2-Dichloroethylene

Under similar circumstances, a saturated compound such as ethane would undergo a *substitution reaction*:

Ethane   Chlorine   Ethyl chloride   Hydrogen chloride

In this reaction, a chlorine atom has been substituted for a hydrogen atom in the ethane molecule.

Two molecular models of ethylene should be constructed, as in fig 2:

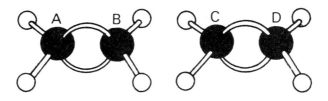

Fig 2

The strained condition of the bonds between A and B and between C and D is a mechanical representation of the chemical instability of ethylene molecules. If one of the spokes is detached from sphere A, the bonds assume a tetrahedral and less strained configuration as in the ethane model, fig 3:

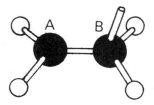

Fig 3

One of the holes drilled in sphere A is now vacant. The second ethylene model should be treated similarly. It will be found that the two molecular models can be joined, giving a 'straight chain' as in the paraffins, see fig 4:

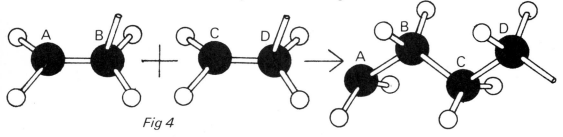

Fig 4

The length of this chain may be increased at will, simulating the formation of a long-chain molecule such as that found in the plastics material, polythene (polyethylene). The linking together of molecules in this way to form larger molecules is known as ***polymerization***:

Poly—(polus, many, Gk.)
Meros—(parts, Gk.)

The product of a polymerization reaction is known as a ***polymer***.

The small unit forming the link in the chain, in this case ethylene, is known as the *monomer.*

3 Types of Polymerization

Addition Polymerization

This occurs when molecules of the monomer, e.g. ethylene, react additively producing long chain-molecules with up to about 20,000 links in the chains.

Ethylene → Polythene

Condensation Polymerization

This type of polymerization usually involves two **different** kinds of monomers.

The molecules of at least one of these monomers contain two or more reactive groups of atoms while the molecules of the other may contain one or more reactive groups. Chemical interaction of these monomers produces a plastics material having either long chain-molecules or a three dimensional cross-linked structure. In either case, molecules of such simple substances as water (H_2O) or hydrogen chloride (HCl) are eliminated. It is in this respect that condensation polymerization differs from addition polymerization.

Example: Urea-formaldehyde **resin***

Urea + Formaldehyde → Urea-formaldehyde resin

Polymer Classification

1 **Natural:** rubber, proteins, cellulose.

2 **Semi-synthetic:** cellulose acetate, nitrocellulose (celluloid).

Both of these polymers are manufactured from a naturally occurring material, cotton linters, by treatment with appropriate chemicals under specific conditions.

3 **Synthetic:** Perspex, Bakelite, polystyrene, nylon.

* The term 'resin' is frequently applied to synthetic plastics materials before they are moulded. They do not, however, have molecular structures identical with those of naturally occurring resins such as rosin and Congo copal.

9

4 Types of Plastics Materials

There are two groups of these materials, **thermoplastic** (thermosoftening) and **thermosetting**.

Thermoplastic materials can be softened and moulded (usually under pressure). Hardening takes place on cooling. It is a purely physical process and chemical hardening does not take place (contrast thermosetting). The cycle of heat-softening, moulding and cooling can be repeated without any significant change in the chemical and physical properties of the material.

Thermoplastics have linear chain structures as shown in fig 5:

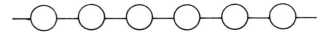

Fig 5

◯ represents the simple repeating unit derived from the molecule of the monomer. These molecular chains are not bound together by rigid chemical bonds. They are linked by weak electrostatic forces known as van der Waals forces. These intermolecular forces can be overcome by comparatively small rises in temperature; the distances between molecular chains increase, and the material softens.

Thermosetting plastics materials differ from thermoplastic materials in the following ways:
1 They can be heat treated only once.
2 They are insoluble in **organic** solvents such as benzene and acetone. The term 'organic' infers that the solvents are **carbon** compounds.
3 They are three-dimensional polymers having rigid 'network structures', see fig 6:

Polymer Structures **Rubbers.** The structure is amorphous (non-crystalline). The molecular chains are coiled and are arranged in a random manner while the intermolecular van der Waals forces are small. Under tension, the molecular chains straighten out and the rubber stretches. When the tension is released, the chains

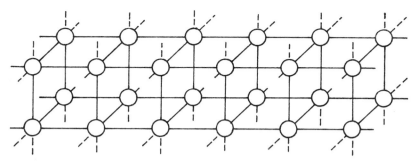

Fig 6

recoil with little interference from neighbouring chains and the rubber contracts.

Thermoplastics. The structure is usually partly amorphous (molecular chains entangled) and partly crystalline (molecular chains aligned). It will be seen later that the degree of alignment of molecular chains and therefore the degree of crystallinity in the structure of thermoplastics can be altered by chemical and physical manipulation. The intermolecular forces in thermoplastics are greater than those in rubber, particularly in the crystalline regions. Hence thermoplastics are more rigid and less elastic than rubber.

Thermosetting plastics. Cross-linking between molecular chains produces a rigid giant molecular structure which is inelastic.

5 Modification of Polymers to produce Desirable Qualities in Plastics

Chemical Modifications

1 It has been seen that ethylene is the basic unit (monomer) in the manufacture of polythene. The replacement of some or all of the hydrogen atoms in the ethylene molecule by other atoms or groups of atoms yields a variety of monomers which have the same carbon 'skeleton'. Consequently, they all undergo addition polymerization yielding plastics of widely

11

varying physical and chemical properties. The following are examples:

(a)

$$n\left(\begin{array}{cc} & \overset{\displaystyle H}{\underset{\displaystyle H-C-H}{|}} \quad H \\ & C = C \\ & \overset{\displaystyle |}{H} \quad \overset{\displaystyle |}{H} \end{array}\right) \longrightarrow \left(\begin{array}{cc} & \overset{\displaystyle H}{\underset{\displaystyle H-C-H}{|}} \quad H \\ & -C- \quad -C- \\ & \overset{\displaystyle |}{H} \quad \overset{\displaystyle |}{H} \end{array}\right)_n$$

Propylene   Polypropylene

The substitution of a methyl group $\left(\begin{array}{c} H \\ | \\ H-C- \\ | \\ H \end{array}\right)$ for a hydrogen atom in the ethylene molecule leads to the production of a stiffer and more heat-resistant plastics than polythene.

(b)

$$n\left(\begin{array}{cc} H & Cl \\ | & | \\ C = C \\ | & | \\ H & H \end{array}\right) \longrightarrow \left(\begin{array}{cc} H & Cl \\ | & | \\ -C-C- \\ | & | \\ H & H \end{array}\right)_n$$

Vinyl chloride   Polyvinyl chloride (PVC)

The presence of a chlorine atom in the monomer renders these plastics self-extinguishable.

(c)

Styrene   Polystyrene

Again, the structure of the styrene monomer is ethylenic.

One hydrogen atom in an ethylene molecule has been replaced by the phenyl (C_6H_5) group of atoms. This group is

represented above by the symbol ⌬. This symbol is an

12

abbreviation for the full structural formula of the phenyl group and is therefore more convenient to use than the formula shown below.

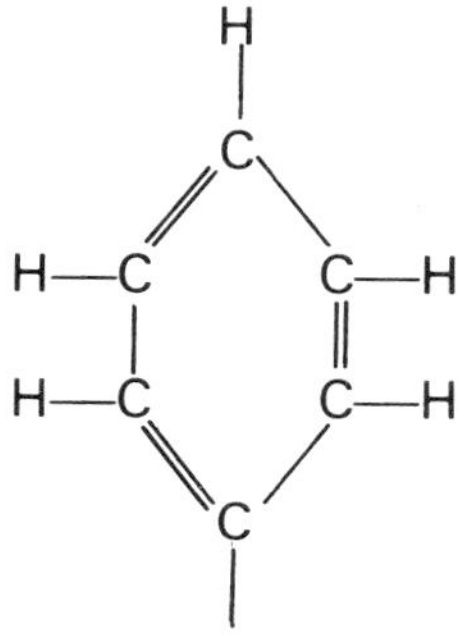

Polystyrene is more brittle than polythene and has a higher density.

$$\text{Density of polystyrene} = 1\cdot05 - 1\cdot07 \text{ g/cm}^3$$
$$\text{Density of polythene} = 0\cdot92 - 0\cdot96 \text{ g/cm}^3$$

Polystyrene dissolves in such organic solvents as aromatic hydrocarbons (benzene) and certain esters (propyl laurate). Polythene does not dissolve in any liquid at room temperature.

(d)

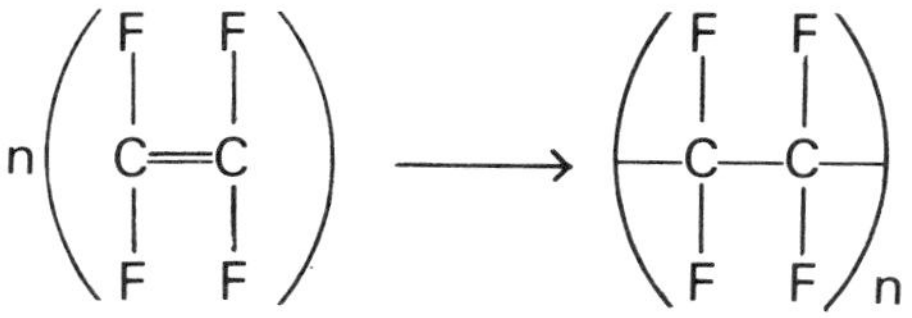

Tetrafluoroethylene   Polytetrafluoroethylene (PTFE)

In the above formulae, the letter F represents one atom of the element fluorine which has chemical properties similar to those of chlorine.

Four hydrogen atoms in the ethylene molecule have been replaced by fluorine atoms. This chemical modification of ethylene produces a polymer (PTFE) having the following useful properties:

(a) High chemical resistance.
(b) A high softening point (327°C)—polythene softens in the range 76–120°C, depending on the method of manufacture.
(c) Non-inflammability.

13

(d) Low coefficient of friction which makes PTFE suitable for non-lubricated bearings.

2 Copolymerization

When two or more different monomers are polymerized the product is known as a **copolymer**. The properties of the copolymer are often more desirable than those of the polymers formed separately from each of the monomers. The following are examples:

(a) **Vinyl chloride/Vinyl acetate copolymer.** A mixture of about 90 parts of vinyl chloride monomer and 10 parts of vinyl acetate monomer, when polymerized, yields a product which combines the toughness of PVC with the heat stability of polyvinyl acetate (PVAC). Since PVAC is soluble in a wide range of organic solvents while PVC is not, it is found that the copolymer is more soluble than PVC. This renders the copolymer more useful when coating techniques are employed. In the molecule of the copolymer there is usually a random arrangement of the different monomer units:

Vinyl chloride   Vinyl acetate

Vinyl chloride/Vinyl acetate copolymer

(b) **Styrene/Butadiene copolymer.** Butadiene has the structure:

Since there are **two** pairs of carbon atoms in the molecule, each linked by a double bond as in ethyl**ene**, the term **diene** is used. There are four carbon atoms in the molecule, as in the paraffin, butane, therefore the hydrocarbon is called 'butadiene'. In the presence of a sodium catalyst,[1] the butadiene polymerizes to produce **buna** rubber. The term **buna** is derived from the first two letters of BUtadiene and from the chemical symbol for sodium, Na.

Butadiene   Buna rubber

Buna rubber is not an attractive commercial material since it has a low tensile strength and it is difficult to process.[2] However, when butadiene is copolymerized with styrene much tougher rubbers are obtained. These rubbers are known as **S**tyrene/**B**utadiene **R**ubbers (SBR) polymers). SBR polymers can be vulcanized and are used in the manufacture of tyres.

The structural formula shown below is not intended to convey any information concerning the proportion of the two monomers used in copolymerization.

SBR polymer

The chemist has introduced the large phenyl group into the copolymer molecule chain to increase the entanglement of adjacent molecular chains. This reduces the degree to which the chains are aligned under tension and prevents crystallization which would adversely affect the elasticity of the polymer.

[1] The term 'catalyst' is defined later in this chapter.
[2] Modern Ziegler catalysts can, however, be used to produce commercially viable polybutadiene rubbers.

15

3 Catalysis

A catalyst is a substance which alters the rate of a chemical reaction without any change in the quantity of the catalyst itself.

A positive catalyst increases the rate of a chemical reaction and is sometimes known as an 'accelerator'. Sodium hydroxide (an alkali) accelerates the formation of phenol-formaldehyde resins.

A substance which improves the performance of a catalyst is known as a promoter. The efficiency of catalysts such as benzoyl peroxide and lauroyl peroxide are increased by the presence of ferrous salts.

Some catalysts reduce the rate of a chemical reaction and are known as negative catalysts.

A substance which reduces the efficiency of a catalyst is called an 'inhibitor'.

Methyl methacrylate monomer polymerizes easily under the catalytic influence of light. The shelf life of the monomer is lengthened by the addition of hydroquinone, which is an inhibitor.

Polythene is made by a high-pressure and a low-pressure process. In the high-pressure process ethylene, which is a gaseous product of the petroleum chemicals industry, is polymerized at a pressure of over 1000 atm. and at a temperature in the range of 200–250°C. Small quantities of oxygen have a catalytic effect. The molecular chains of the product have the following characteristics:

(a) Comparatively short length.
(b) They are branched, and branching may be excessive thus preventing the development of crystalline areas (crystallites).
(c) Random orientation.

These characteristics prevent molecular close-packing, with the result that the polythene has low density (0·92–0·94 g/cm^3) but greater flexibility. It also has a low softening point (about 75°C).

In the low-pressure process, ethylene gas is passed into a suspension of a Ziegler catalyst in a suitable liquid hydrocarbon at 60°C and at a pressure of 1 atm. Professor

Ziegler introduced this method of making polythene in 1953, in Germany. There are several Ziegler catalysts but one of them is aluminium triethyl $(Al(C_2H_5)_3)$. Titanium tetrachloride $(TiCl_4)$ is used as a promoter. The molecular chains of this product have the following characteristics:

(a) Longer length than those produced in the high-pressure process.
(b) Low degree of branching.
(c) High degree of alignment.

Characteristics (b) and (c) allow close-packing of molecular chains and the formation of crystalline regions known as crystallites (fig 7).

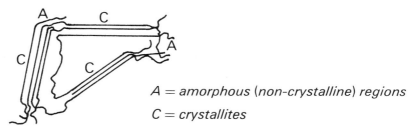

A = amorphous (non-crystalline) regions

C = crystallites

Fig 7

It should be noted that the crystallites are randomly orientated.

A catalyst which produces molecular alignment is called a stereo-specific catalyst. The polythene produced by the low-pressure process has a higher density $(0.94–0.96 \text{ g/cm}^3)$, greater rigidity, higher tensile strength and a higher softening point (about 120°C) than the product of the high-pressure process.

4 Cold Drawing

If a polymer such as nylon is subjected to a carefully controlled stress, necking occurs (fig 8).

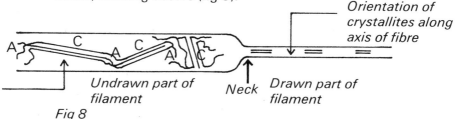

Orientation of crystallites along axis of fibre

Random orientation of crystallites

Undrawn part of filament

Neck

Drawn part of filament

Fig 8

The effect of stretching the polymer is to orientate the crystallites in the direction of stress and to bring into alignment those molecular chains which were originally in the amorphous areas of the polymer. This treatment produces a fibre of good tensile qualities in the direction that matters, viz. along its axis.

Physical Modification

1 *Plasticizers* are materials which are added to polymers to make them less brittle and to lower their softening temperatures. Plasticizers, which are usually liquids but may be solids, must be carefully selected. The plasticizer should be a good solvent for the polymer, i.e. the plasticizer and the polymer should be compatible.

The plasticizer should also have a low vapour pressure in order to increase its permanence. High molecular weight organic compounds such as dibutyl phthalate fulfil these conditions. Dibutyl phthalate is added to methyl methacrylate monomer. In this way, polymerization produces a more flexible material than ordinary 'Perspex'.

2 *Fillers* are inert solids which are added to resins or polymers to modify the properties of the material. Thus wood flour, alpha cellulose and cotton, glass or synthetic fibres reduce brittleness and increase impact strength. Asbestos fibres and mineral powders improve heat- and/or chemical-resistance, while mica flake improves heat-resistance and chemical properties. Fillers are almost invariably cheaper than the pure resin, and their use lowers the cost of the end product. Fillers used mainly to lower cost with little change in other properties are termed 'extenders'.

Experimental Work

It is recommended that the student should follow up the reading of this chapter by carrying out experiments in Polymer Chemistry under the guidance of the Chemistry Department of his school or college. Suitable experiments are provided in *Organic Chemistry through Experiment*, published by Mills & Boon. Alternatively, the Griffin Polymer Kit (S 74–750) is obtainable complete with experimental instructions from Messrs Griffin & George Ltd.

Section II

THE MAKING OF PLASTICS

1 Sources, Manufacture, Main Forms

Source of Materials

The basic raw materials for plastics are coal and oil, and many plastics materials today are produced from oil.

When crude oil is refined the process produces several layers of density from the very heavy tar end at the bottom to the light petroleum end at the top. Manufacturers of plastics purchase their raw material from the refinery. It is the light petroleum end which is of interest to them because they are able to feed this into a cracker plant and separate the monomers (e.g. ethylene, propylene, butadiene) which are essential to their process. After the cracking process the monomers are piped to various plants which are concerned with the manufacture of different types of plastics.

Research

The research chemists produce new materials in small quantities under laboratory conditions, some of which go on to pilot production for proving purposes and thence into the full production phase.

Manufacture

Complex and expensive plant is used to produce plastics materials, but the job they do is basically a simple one. The final process in manufacture involves polymerization, that is single molecules—monomers—are made to link up—polymerize—to form long chains called **polymers**.

Forms of Plastics

From the manufacturing processes plastics emerge in a large variety of forms—fine powders, emulsions, resins, viscous fluids, pellets, granules and cubes.

Basic Categories

There are two basic types of plastics:
1 Those which harden under heat, set and cannot be reconstituted to their former state—**thermosetting materials**
2 Those which soften when heated and set upon cooling and will do this many times over—**thermoplastic materials**

20

Thermosetting Materials

1 **Phenolic Resin**: Electrical switch and plug covers, bottle and container tops, door handles, ash trays, telephone handsets.

2 **Urea-formaldehyde Resin**: Adhesives, wood impregnation, surface coating of metal, laminating timber.

3 **Melamine-formaldehyde Resin**: Moulding, decorative laminated sheet, tableware, cloth and paper impregnation.

4 **Polyester Resin**: Usually reinforced with glass fibres and used in boat and car body production, container ware, ducting, 'radomes' for aircraft, building techniques.

5 **Epoxy Resin (Epoxide)**: Surface coating, casting, adhesives, laminating and moulding.

6 **Polyurethanes**: In foamed forms polyester polyurethane provides buoyancy in boat hulls, reinforcements and insulation; polyether polyurethane is used for artificial sponge production and for cushioning.

Thermoplastic Materials

1 **Polyethylene**: Household goods, insulation of electrical wires, children's toys, bag and film for packing.

2 **Polypropylene**: Laboratory ware, tableware, toilet seats, heels for ladies shoes, chair seats, filaments for brooms and brushes, items of plumbing.

3 **Polyvinyl Chloride (PVC)**: Water and drain piping, packaging, rainwear, coating of fabrics, floor tiles, gramophone records, foam cushioning, roofing, hose pipes for the garden.

4 **Polystyrene**: Household containers, toys, refrigerator parts, expanded insulating material, packaging.

5 **Polymethyl Methacrylate (Acrylics or 'Perspex')**: Light fittings, skylights, aircraft canopies and windows, advertising signs, baths, motor-car rear-lamp lenses, watch and clock 'glasses'.

6 **Polytetrafluoroethylene (PTFE)**: Bearing surfaces, high-quality electrical insulation, lining pumps and pipes carrying highly reactive chemicals, non-stick surfaces in holloware.

7 **Cellulose Acetate**: Fountain-pen cases, packaging film, lacquer manufacture, fibre production, cine and still films.

8 **Nylon**: Stockings, clothing, combs, toothbrush filaments, bearings.

9 **Acrylonitrile-Butadiene-Styrene (ABS)**: Injection moulding and vacuum forming processes for boat hulls and document cases.

21

2 The Processing of Plastics Materials

When plastics materials leave the manufacturers as fine powders, emulsions, resins, viscous fluids, pellets, granules or cubes, these various forms are taken and processed to render them a usable, workable material, from which an article or component may be produced with the help of machinery.

Plate 1   *Extrusion in industry*
Courtesy of ICI Plastics Division

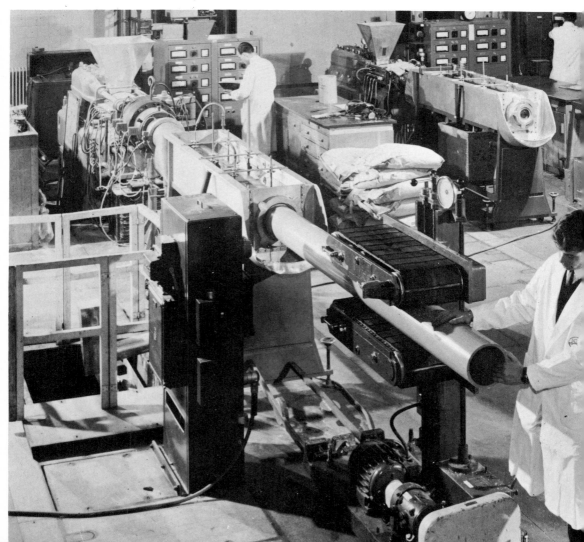

The main processes by which these plastics materials are converted to forms which are familiar to us are as follows:

Compression Moulding

A raw thermosetting plastics in powder form is placed in a mould and subjected to heat and pressure. Handles for electric irons, toilet seats and screw-on bottle tops are three of the many articles produced by this process.

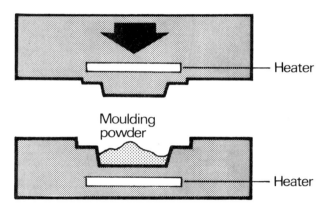

Fig 9

Injection Moulding

A thermoplastic material in granule form is softened by heat and injected into a mould form where it subsequently hardens. A vast number of components are so moulded, including buckets, dustbins, telephones, flowerpots, golf tees, ball pens.

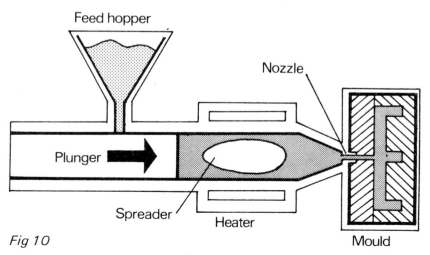

Fig 10

23

Extrusion
A thermoplastic material in granule form is softened by heat and squeezed through a shaped orifice to produce lengths of material. Garden hose, curtain rails, insulated wire, rainwater piping are good examples of extrusion.

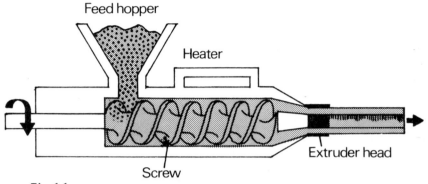

Fig 11

Blow Moulding
Air is blown into a heated thermoplastic tube, thus causing it to inflate and take up the shape of the split mould. This process is widely used in the production of bottles, water carriers, toys.

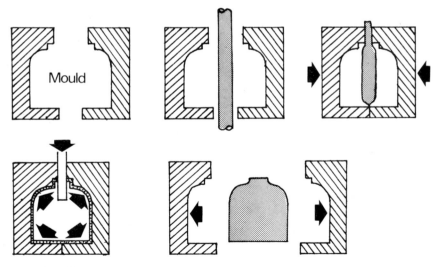

Fig 12

Vacuum Forming
A thermoplastic sheet is clamped in a frame, heated and then drawn down onto a mould form by vacuum pressure.

24

Refrigerator linings, egg boxes and boat hulls can be produced by this technique.

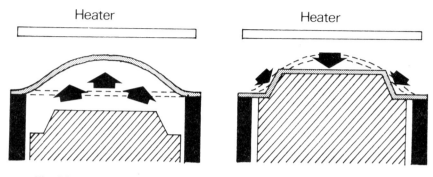

Fig 13

Calendering

A heated mass of thermoplastic material is passed between heated rollers and finally cooled rollers, emerging as flat film or sheet. Material for raincoats, shampoo sachets, covers for motor cars and waterproofing in building is made in this way.

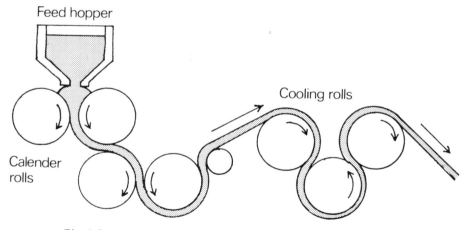

Fig 14

Spreading

A plastics paste, usually PVC, is poured onto a moving length of material, e.g. cotton or carpeting, spread out to a uniform thickness and cured finally with heat. Such coated fabrics are used in garment production, car seating, for covering items such as travelling cases and manufacturing Vinyl washable wallpaper.

25

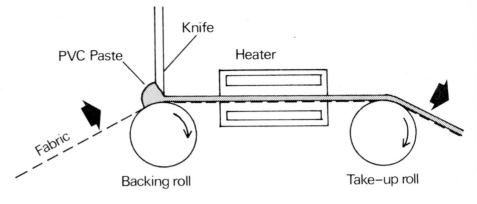

Fig 15

Laminating

Layers of thermosetting resin-impregnated paper or cloth are subjected to heat and pressure to form a single piece. Such materials as decorative laminate sheet for table tops, circuit boards for electrical and electronic circuits, resin-impregnated veneers for pressings are produced in this manner.

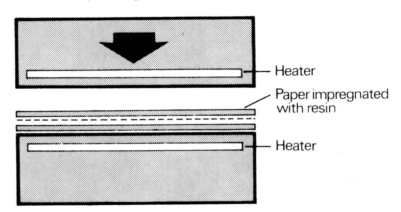

Fig 16

Figs 9 to 16 are based on those in Booklet 7, 'Shaping Plastics', produced by BP Educational Service.

Section III

PRACTICAL WORK– FIRST STEPS

1 Polyethylene Tiles

It would be very difficult to select a more suitable material than polyethylene at the commencement of work with plastics. It has been suggested that polyethylene started a new era in plastics, and when one looks at the progress made in the last twenty-five years in the plastics industry and the public's acceptance of the material, this is certainly no overstatement.

Many new materials and industrial processes came into existence because of the demands of World War II, and when, in 1945, the war came to an end, the demand for these materials and industries suddenly came to a halt. Plastics, which had played such an important and vital part in the war effort, required a market if their existence was to continue. The manufacturers decided that if they could produce equipment to meet the demands and high standards required by the armed services, then they could equally produce, with assured success, articles for the domestic market.

Unfortunately, their enthusiasm was somewhat dampened when, in the early years after 1945, plastics failed to perform adequately and the public quickly decided that the word 'plastic' was synonymous with such phrases as 'cheap-looking' and 'short-lived'; plastics were considered to be a bad buy!

In 1948, with considerable trepidation and history against them, manufacturers launched the first plastics holloware—washing-up bowls moulded in polyethylene. In due course these articles were seen to be superior to their enamelled-metal counterparts, and plastics were on the way to becoming a respectable material. With this success established, the industry began to branch out into the moulding of other household articles which exploited the properties of the material. Careful consideration of the questions of design and utility at this time immensely improved the status of plastics materials.

In the nine years from 1948 to 1957, the amount of polyethylene used in the production of household wares rose from 1 per cent to 40 per cent. Today, we all accept the polyethylene bowl, bucket, bag and bottle, and perhaps we wonder how we ever managed without them. Polyethylene has done much for the name 'plastic'.

The Types of Polyethylene

There are two different types produced: high-density polyethylene is manufactured using a low-pressure process which is carried out in solution; low-density polyethylene is produced by a high-pressure process carried out in a gas phase (see Section 1,5).

Forms of Polyethylene

Once the raw material has been produced in granule or powder form by one of the above processes, it is converted into film and sheet by extrusion or calendering. The natural colour of polyethylene is a milky-white, but it can be processed further into a clear translucent state or it can be pigmented giving a wide range of colours.

Making Polyethylene Tiles

In the production of tiles, a finely ground powder of the low-density variety is used.

It is a versatile material, for it can be used for several other activities as will be seen later on. It is also an easy material to work with and everyone is assured of success. Perhaps the Art Department will be interested in helping with the design of tiles.

Material: Low-density polyethylene powder.

Equipment: Metal trays, grease or mineral oil, spreaders/levellers, table knives, pliers, scissors, an oven.

Procedure (see fig 17)
1 The metal trays shown can be made up out of tinplate in the school metalwork workshop and, with the size of 115 mm square, a tile of 100 mm can be made once trimming has been done.
2 Lightly grease or oil the inside of the tray.
3 Charge the mould with powder and level off, checking that the corners are filled as well as the rest of the mould.
4 Place in the oven set to operate at approx. 180°C.
5 Leave for a few minutes for the powder to fuse.

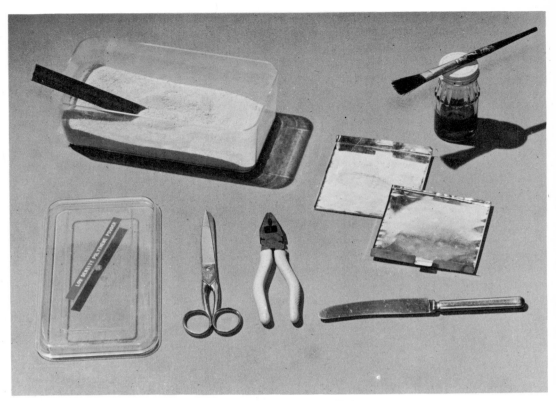

Plate 2   *Equipment for polyethylene tiles*

6 Remove from the oven and allow it to cool for approximately one minute.

7 Quench carefully on the underside of the mould.

8 Ease the tile from the mould with a table knife.

9 Carefully trim the tile, using, if possible, a guillotine or a card cutter.

10 If a decorated tile is required, cut up the first tile made to make a pattern. Prepare another mould, charge with powder and then place the pattern on it. Fuse the powder and the pattern to form a uniform flat mass.

Recent experimental work has shown that it is possible to cast polyethylene tiles immediately onto hardboard with several advantages. This requires that the small tinplate trays be made up with sides of approximately 8 mm—high enough to accept the square of hardboard *plus* the layer of polyethylene

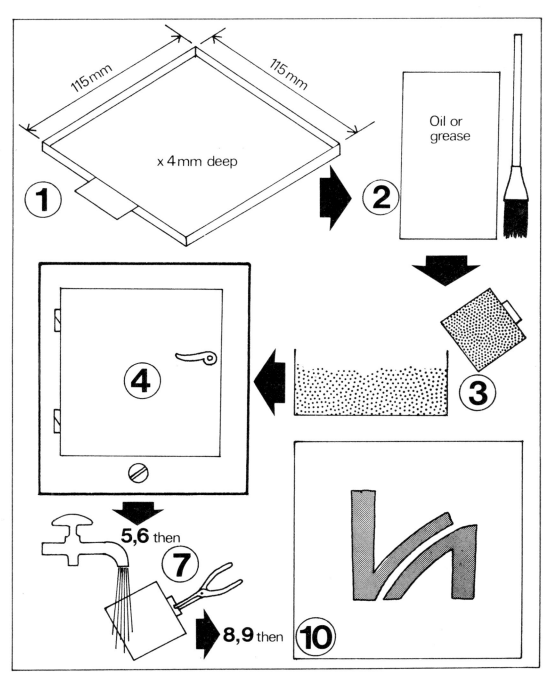

1 115 mm × 115 mm × 4 mm deep

2 Oil or grease

3

4

5,6 then **7**

8,9 then **10**

Fig 17

powder. The mould need only be oiled or greased around the sides where the plastics material will come into contact. The tile mould with hardboard and powder can then be placed in the oven and heated. The powder will tend to settle as before but will also separate, so additional 'topping up' will be necessary, but eventually a perfectly flat smooth surface will be obtained. Throughout, the hardboard remains unaffected by the temperature of approximately 180°C and best adhesion will be obtained if the rough side of the hardboard is positioned so it lies in contact with the plastics material. The main advantage to this technique is that the problem of obtaining good adhesion between tile and undersurface by way of one of the adhesives mentioned does not arise, and it is envisaged that larger areas could be so produced and then the hardboard stuck down to another surface in turn.

Safety Notes

1 Respect a hot oven and use it with care and common sense.
2 Take care in the use of the table knife and do not choose a really sharp one.
3 A guillotine or card cutter should only be used under the supervision of the teacher.

Some observations

Now that a polyethylene tile has been produced, the following questions may be considered:

1 What does the tile feel like?
2 Has polyethylene any smell?
3 How does the tile react with water?
4 Is it flexible and supple or stiff?
5 What proof is there that polyethylene is a thermoplastics material?

Some Applications of Tiles

Here are just some uses for tiles which can be the subject of individual or group work:

1 Small mats
2 Table tops
3 Art murals

The small mats can be mounted on hardboard to give the tile rigidity.

Table tops can be made up on plywood or blockboard and the underframe designed and made up in the woodwork or metalwork shops.

Plate 3　　　　　　*The Griffin Oven*

Murals and art work can be mounted on similar surfaces, framed and hung. This aspect of tile design and production could present interesting group work.

Adhesives for Tile Fixing

Polyethylene is not an easy material to stick down to any surface and no adhesive can be guaranteed to give a satisfactory 100 per cent bond. This is due to the nature of polyethylene itself; it is naturally a greasy or waxy material, it is resistant to many alkalis and dilute acids, and adhesives find it difficult to penetrate its surface. However, reasonable bonds have been obtained with two adhesives and there may well be others to compare as favourably.

Araldite AZ 107 with Hardener HZ 107: This is an epoxy resin of low viscosity; the hardener is also of low viscosity. When they are mixed together in the correct proportions they produce an adhesive which is cold-setting and which will bond a variety of materials to a variety of surfaces.

Evostik Impact Adhesive: This is an adhesive used with plastics laminates and if the manufacturer's directions are followed it will produce quite a good bond.

Plate 4   *A selection of tile designs*

With both these adhesives it is essential to pay attention to the following points:

1 Thoroughly de-grease the back of the tile, using 1 : 1 : 1-trichloroethane.
2 Roughen the surface with coarse glass paper.
3 Roughen the surface to which the tile is to be stuck also with coarse glass paper.
4 Apply the adhesive as directed in the instructions.

2 Resin Moulding

Introduction

This process is concerned with the shaping of articles by pouring a liquid plastics into a mould. Work of this kind may have been covered already in the workshop under the heading of foundry practice and the procedure is very much the same. The plastics sets into a hard mass when it has been treated with special chemical agents. The plastics material used is called polyester resin and it is a thermosetting material. (It is also known as an unsaturated resin, the saturated type being used in the production of cloth fibres. See Section I.)

Polyester resins are very easy to handle for they can be rapidly cured at room temperatures without special equipment. They can be coloured by using special pigments or ordinary powder paints. The finished product made of the resin is fire- and corrosion-resistant and, if it is strengthened by laminating, it can be extremely tough and resilient. Reference to BS 4163: 1968, entitled 'Safety Precautions—Glass Fibre Work', gives safety instructions for polyester resins.

Types of Resins

Polyester resins are used today in a wide variety of situations and hence formulations of resin are manufactured to meet almost any requirement. Resins can be slow-curing or fast-curing; rigid or flexible when hardened off; translucent or opaque; resistant to ultra-violet light if the fabrication is to be outside; impact-resistant; specifically suited to thin-section work; suitable for thick mouldings.

35

Other Materials needed .

The resin if left by itself will not cure and harden off. Heat must be applied to produce polymerization; a quantity of resin placed in a metal mould will harden if it is heated in an oven. However, for the majority of work in which polyester resin is used, certain chemicals are added to the resin to produce polymerization. The first one is known as a 'catalyst', which is a chemical compound added in a small proportion to a monomer to speed up polymerization; the second is known as an accelerator, activator or promoter and this is a substance added to a 'catalyst' to increase its efficiency. By adding a catalyst and an accelerator to polyester resin, polymerization commences and the change from liquid to solid is produced in due time. An intermediate point in this process is known as the 'gel stage' and at this point the resin assumes a jelly-like state where it is quite flexible and tough.

The 'catalyst' is usually methyl ethyl ketone peroxide (MEKP) and the accelerator, activator or promoter is cobalt naphthenate. Both of these materials must be kept well apart in storage and in use as together they form an explosive mixture. It is advisable to follow very closely the supplier's or manufacturer's instructions on the mixing of polyester resin, 'catalyst' and accelerator. Most resins are sold now in a pre-activated state; that is accelerator, activator or promoter has been added, and only catalyst is needed to produce polymerisation.

Storage of Materials

Resins remain stable for upwards of twelve months if the surrounding temperature is kept to 20°C (68°F) or lower, although too low a temperature can be just as harmful as too high a temperature. 'Catalyst' should be kept in sealed containers, in the dark if the container is glass or plastics, and at a temperature not above 20°C (68°F). Accelerator, activator or promoter has a shelf life of about six months if it is stored in sealed containers in the dark—if the type of container requires it—at a temperature not above 20°C (68°F).

Some local education authorities have regulations regarding the storage and use of resins and ancillary materials.

Making Resin Mouldings

Materials: Polyester resin (lay-up grade), 'catalyst', accelerator, pigment or powder paint.

Equipment: Tobacco or polish tins, candle wax, carving tools of various sizes and designs, felt or fibre-tip pens, bunsen burners, 100 c³ beakers or waxed cups, glass stirring rods, a good supply of paper towels, acetone.

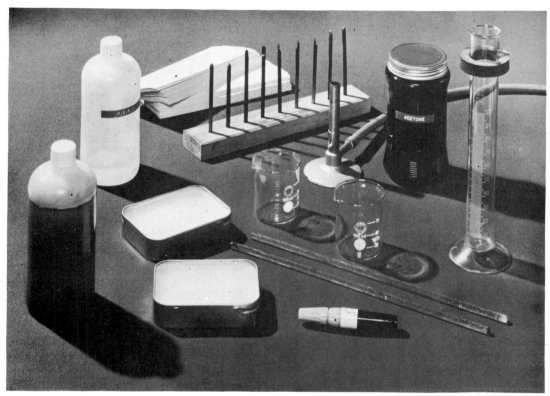

Plate 5　　　　　　　*Equipment for resin mouldings*

Procedure (see fig 18)

1 The tins should be filled with wax to about 5 mm from the top edge; if the wax is lower it will make carving the design more difficult.

2 Allow the wax to harden thoroughly before use, then draw on the design using a felt or fibre-tip pen.

3 A simple design made up of straight lines is shown and the carving can be started using any sort of carving tool that cuts into the wax satisfactorily and gives the desired shape. Special

37

care should be taken to see that adequate draught is present on all vertical surfaces of the mould, for without it withdrawal of the casting will prove difficult. The depth of the cut should be about 4 mm.

4 With the mould completed, sufficient resin, 'catalyst' and pigment or powder paint are mixed. It is worth mentioning here that 100 c³ of resin will be sufficient for about six 50 mm square mouldings.

5 Add accelerator in the correct proportion, if the resin has not been pre-activated. Stir well, then fill the mould cavity.

6 When the gel stage is reached, the moulding can be eased from its mould carefully. Trim off any flashing around the edge, using scissors. Place the moulding on a polyethylene sheet.

7 Additional resin, 'catalyst' and an alternative colour can be mixed and the accelerator added.

8 The resin can then be poured into various cavities in the casting in order to produce a two-coloured effect. More than two colours may, of course, be used if desired. The moulding should be left to cure completely.

9 All containers and equipment used in resin moulding should be cleaned promptly and this is best done with paper towels. The containers can finally be wiped out with some acetone on a cloth.

Safety Notes

1 The 'catalyst' and accelerator **must** be stored and always kept **well apart** when in use as they form an explosive mixture.

2 All traces of resin, 'catalyst' and accelerator should be removed promptly from the skin. Wash with warm water and soap.

3 Any person prone to skin rashes should wear rubber gloves for this work.

4 Use acetone in a well-ventilated area and keep it, and cloths associated with it, well away from naked lights.

5 Resin mouldings are brittle—handle them with care !

Some Observations

The mould can be used as many times as required. A moulding may be produced for experimental work on the following lines:

1 What does polyester resin feel like ?
2 Has it any particular smell ?
3 How does it react with water ?

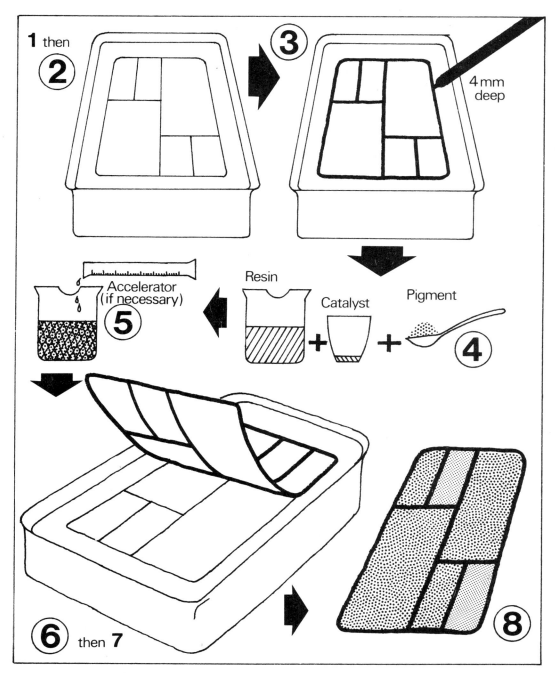

Fig 18

4 Can it be bent?
5 How does it behave after being heated in an oven?
6 How does it burn? Hold a small piece in a Bunsen flame.

Some Applications This work is essentially artistic and decorative. The list shows
some possible uses for castings:
1 Pendants, brooches, earrings and jewellery in general
2 Decorations for other craft work—table lamps, lamp shades,
boxes
3 Murals

Important note: Some resins are sold in a pre-activated
condition—that is, accelerator, activator or promoter has been
added and all that is needed to initiate polymerization is the
'catalyst'.

Plate 6 *A display of resin mouldings made in school*

Resin Moulding with Reinforcement

Introduction

In this work, which is an extension of the small resin moulding covered in the preceding section, the size of the work demands that some degree of reinforcement be introduced during the manufacture.

Commercial everyday examples of this reinforcing technique are many: glass-fibre car bodies and boat hulls are seen advertised and in use; cabs of lorries and trucks and caravan shells are made in resin reinforced with glass fibre rather than the conventional sheet metal; even trays to carry crockery and glasses are made using this material and technique. All these and many more articles are made from polyester resin, which is a very brittle material, reinforced with glass fibre. A check on the inside surfaces of a reinforced resin fabrication would show many thousands of fibres locked in the resin, all crossing one another and giving the resin strength.

Plate 7   *The body of the Lotus Elan is a glass reinforced plastics moulding.*

Courtesy of Lotus Cars

Reinforcements   Glass-fibre reinforcement has very wide applications in plastics today and it is used to strengthen polyesters, epoxies, phenolics, melamines and polyamides (nylon). Glass fibres or filaments are immensely strong and when combined with plastics offer a high impact resistance. Once embedded in the material they are chemically stable and will not rot, oxidize or deteriorate in any way.

During its application, glass fibre is unaffected by the 'wetting' it receives from whatever plastic surrounds it and it does not stretch or go out of shape. However, too vigorous pulling or smoothing out can disturb the fibres. Finally, it is heat-stable up to 540°C (1000°F) and therefore offers considerable resistance to fire.

Glass fibre is manufactured in a variety of forms, all suited to applications involving plastics:

Glass-fibre mat   Woven fabrics
Fine-strand finishing mat   Unidirectional fabric
Woven tape   Chopped strands
Woven roving   Milled fibre

The mattings are probably the most widely used for building up and reinforcing resins, being much cheaper to produce than the woven materials. Rovings, cloths and tape are the strongest but they are about 50 per cent dearer than the mat. Unidirectional fabric is used where strength is required in one direction only. Milled fibre and chopped strands are the least strong and are used as a filling or a building-up material.

The mat is sold by weight per square foot, e.g. 1 oz, $1\frac{1}{2}$ oz, 2 oz, and it is usually manufactured in rolls. The cloth and fabric are produced in a variety of weaves and patterns and are sold by the yard. The chopped strand and milled fibre are sold by weight.

Finally, glass fibre has a surface treatment or finish and it is important to obtain the correct surface treatment for the proposed work. Some glass fibre is treated for easy saturation; some is coated with silicone which would repel saturation. It is quite simple to judge when complete saturation occurs, for the matting tends to merge with the resin and takes on the colour and translucence of the resin.

42

Making Reinforced Mouldings

Materials: Polyester resin (gelcoat and lay-up grades), 'catalyst', accelerator, pigment or powder paint, glass-fibre mat—the weight will depend on the size of the moulding or whatever is available.

Equipment: Large trays which can be made up in the metalwork shop, wax (the microcrystalline type is best), carving tools to suit the proposed design, felt or fibre-tip pens, bunsen burners or brazing hearth torches, large beakers or waxed cups, glass stirring rods, a good supply of paper towels, cheap paint brushes, scissors, acetone.

Procedure (see fig 19)

1 A large tray of wax is prepared. Typical trays might measure 450 mm square, 600 × 300 mm and 1800 × 900 mm with

Plate 8   *A selection of samples of glass matting*

Plate 9   *Students at work carving a large resin lay-up mould*

the wax about 10 mm deep. To fill the trays, melt the wax in tins over a very low flame and, having poured it into the trays, smooth out with bunsen burners or brazing torches taking care the flash point of the wax is not reached.

2 Draw on the design with the pens.
3 Carve the design, remembering that all lines and shapes will need scaling up in comparison with the previous smaller work.
4 Prepare the glass-fibre mat. Note that where pieces have to be joined, an overlap must be made of about 25 mm. Allow the matting to overlap the edges of the mould all round.
5 The mould should now be brushed over with a light mineral oil—this is the mould-release agent which is needed with microcrystalline wax.
6 A quantity of gelcoat resin, 'catalyst' and pigment can be mixed, noting the quantities of each of these at this stage, especially the amount of pigment used. This will assist in subsequent mixes and colour matching. It should be noted that 453 g of resin will cover an area of 450 mm × 450 mm to a thickness of about 3 mm.
7 Add the accelerator, if necessary, and give the complete mould a thin gel coat. This is a layer of no more than 3 mm which is left to gel over the next few minutes.
8 Mix lay-up resin, 'catalyst' and pigment sufficient for laminating the glass matting. About 200 g of resin should be sufficient to wet out this area of resin and glass matting.
9 Add the accelerator if necessary.
10 Brush some of this lay-up resin over the gel coat which should now be tacky, place on the previously prepared glass-fibre matting, pressing it down with gentle finger and palm of the hand pressure so that it makes good contact with the gel coat. If the matting is in sections, overlap the pieces as previously mentioned.
11 Work the resin through the mat completely, using the brushes with a dabbing or stippling action. *Do not brush* in the accepted way as this will disturb the mat form and pull away stray fibres. Add more lay-up resin to areas of the mat which look dry.
12 As soon as the resin is dry to the touch, remove the casting from the mould, easing it away from the edges and working towards the centre.
13 Trim off the edges to the required size as soon as possible.
14 Place the casting on a flat surface to cure fully.
15 The finished article may be framed as shown in plate 10. Such mouldings can be used for screens, murals and partitions and will look more attractive if lighting can be arranged behind.

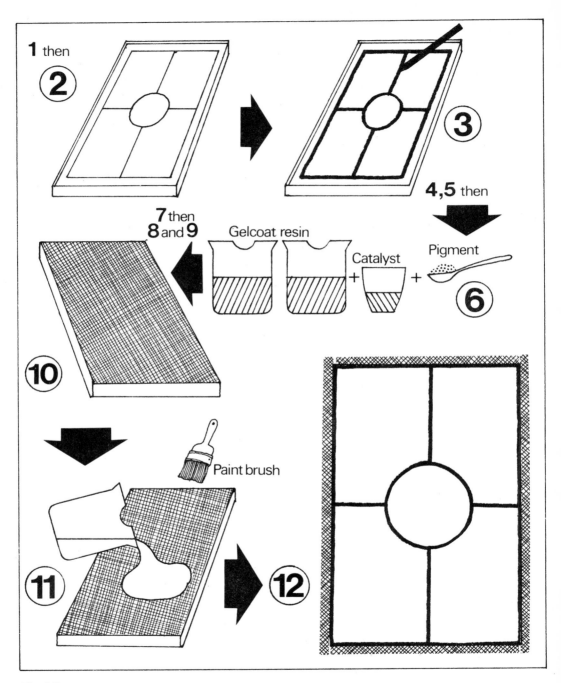

1 then
2

3

4,5 then

7 then
8 and **9**

Gelcoat resin

Catalyst

Pigment

6

10

Paint brush

11

12

Fig 19

Plate 10 *The resin moulding finished and framed*

Safety Notes
These are as for Resin Moulding with one addition:
Glass matting must be handled carefully, for minute fibres can
penetrate the skin and cause irritation. If any person is
susceptible, protective gloves should be worn.

Some Observations
With some offcuts from the casting, try out the following tests:
1 What is the result when the reinforced resin is cut, say with
tin snips?
2 Can it be bent at all?
3 What does the section through the moulding look like?
4 How does a small piece burn in a Bunsen flame?
5 What use does the gel coat serve?

Some Applications   Here are just a few of the many various applications of this work:

1 Decorative wall panels and reliefs
2 Lamp shades
3 Canoe and boat building
4 Sculpture
5 Modelling

Important note: Some resins are sold in a pre-activated condition—that is, accelerator, activator or promoter has been added, and all that is needed to initiate polymerization is the 'catalyst'.

3 Sintering and Casting

Introduction   These two processes are considered together because they have much in common with each other: both require the use of thermoplastic materials; moulds are used to shape the material; an oven is required for both techniques.

By definition, sintering is a process where powder or granules are held at just below their melting point for a short time; the particles of the material become fused or stuck together but not melted. Where granules are concerned, they should not lose their identity but they should remain recognizable as separate pieces of plastics material joined to one another.

Casting employs a fine powdered form of thermoplastics material which is completely melted to mould articles. As the powder heats up, fuses and finally goes into the molten state, it takes up the details and shape of a metal mould. When the melting is complete, the mould can be cooled and the component removed.

1 Sintering   This technique is used in the production of certain thermoplastics, converting their powdered or granular form into standard bar, rod, tube or sleeving. Usually a first step is to preform the material into a shape which is subsequently

48

sintered. The preform is then heated to the temperature at which the individual particles in the preform soften and bond together.

In a school situation, sintering can be used to produce some interesting two-dimensional and three-dimensional shapes, and these could have considerable appeal to the artist. Sintering can, incidentally, use up odd amounts of granules left over from injection moulding.

Making Sinterings   **Material:** Pellets or granules as used for injection moulding in polystyrene and polyethylene.

Equipment: Tin trays—square, triangular, rectangular, hexagonal in shape—oil or grease, pliers, an oven, adhesive.

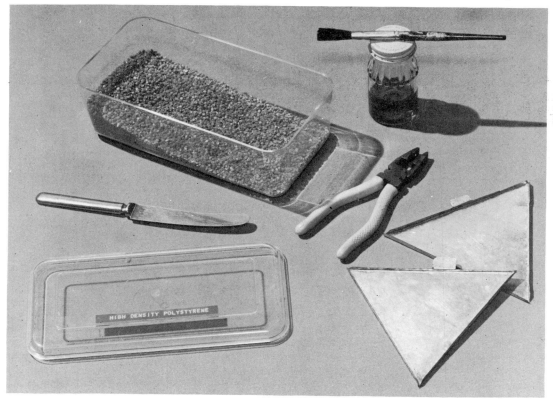

Plate 11   *Equipment for sintering*

Procedure (see fig 20)

1 The metal trays could be the same ones as were used in Tile Making; new shapes may be added so that a variety of two-dimensional and three-dimensional structures can be built.
2 Lightly oil or grease the tray.
3 Charge the tray with granules—the amount of material used will depend on whether a loose-textured form or a denser one is desired.
4 Place in the oven—set to operate at 160–180°C.
5 As the granules heat up, the stress produced in them during the manufacturer's extrusion process is released and their shape can change from a cylindrical to saucer-like form. The granules may also tend to rise up in the tray, but gentle downward pressure with the blade of a knife will overcome this.
6 When the material has sintered to the required degree, remove the tray from the oven and allow to cool for a minute or so in air.
7 Quench and ease the shape carefully from the mould.
8 When a number of shapes have been produced, construction of designs or three-dimensional forms can commence.

Safety Notes

1 Respect a hot oven and use it with care and common sense.
2 Take care when handling trays or materials that have just been removed from the oven.

Some Observations

1 Take a close look at one of the shapes and compare the form of the extrusions after sintering with those not sintered.
2 If there is a spare shape, test its strength.
3 If you are using several different thermoplastic materials, compare their physical properties by bending, dropping and twisting.

Some Applications   This work lends itself to individual or group work and here are some possible approaches:

1 Two-dimensional or three-dimensional artistic murals
2 Three-dimensional geometrical models
3 Lamp shades

50

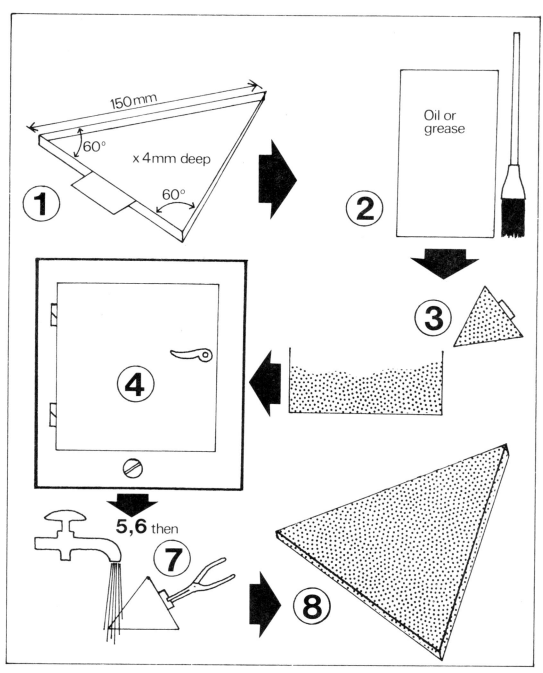

Fig 20

Adhesives

A solvent can be used as a basis for an adhesive, for if it is applied to the edges to be joined, it will dissolve the material and with pressure it will effect a good bond. A few granules of the material that is being joined can be added to the solvent to provide a thicker adhesive and one with greater holding power. This can work particularly well with polystyrene when it is added to 1 : 1 : 1-trichloroethane. Sinterings in polyethylene should be joined using the adhesives mentioned in the section on Polyethylene Tiles. Remember to use 1 : 1 : 1-trichloroethane in a well-ventilated atmosphere.

2 Casting

Co-operation will be needed with the metalwork shop because the important piece of apparatus is the metal mould. Moulds can be made up in steel or brass, the latter being particularly suitable because a good finish can be produced on it relatively easily. The finish on the mould surfaces is very important, for any blemishes or marks will be reproduced on the casting. Time spent on seeing that turning tools are sharp and that emery clothing is carried out correctly will pay dividends. The resulting casting will have a good appearance and finish. An interesting aspect of this type of casting is that the end product is stress-free as no external pressures have been exerted on the material in shaping it.

Making Castings

Material: Low-density polyethylene powder as was used in tile making.

Equipment: Moulds, oil or grease, a tray on which to place moulds, an oven.

Procedure (see fig 21)
1 As indicated, the most important piece of equipment is the mould—success depends on its shape, accuracy and finish. Some moulds are shown for casting gear wheels, and push-on table feet. These can all be made by boys in the metalwork shop, so a degree of co-operation with this part of the school is essential.
2 Prepare the mould with a very light coating of oil or grease.
3 Charge the mould and press the powder down well, especially into corners or any small detail that the mould may have.
4 Place the mould on a tray and then in the oven set to operate at 180°C.

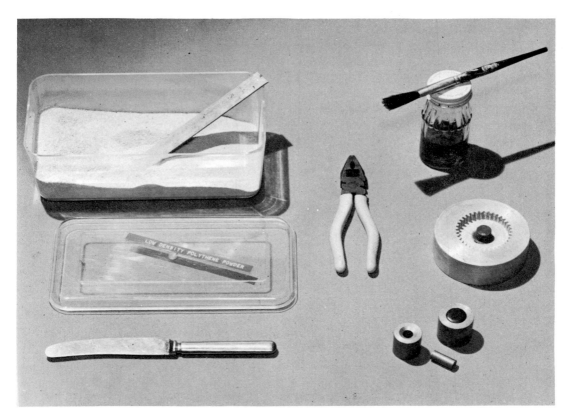

Plate 12　　　　　　　　*Equipment for casting*

5 As the mould and powder heat up, the level of the powder will sink as air escapes, so a 'topping-up' step has to be introduced when the settling down is apparent.
6 Continue heating until fusion is complete.
7 Remove the mould from the oven and allow to cool in air for several minutes, depending on the size of the mould.
8 Now quench and eject or remove the component from the mould.

The component produced by this process is similar to one that would be normally produced by the injection moulding process, but this has been done with much simpler equipment.

53

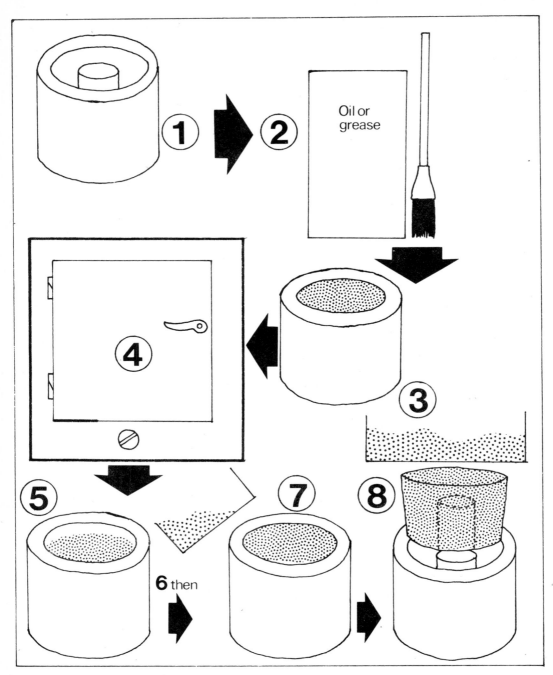

Oil or grease

Fig 21

Safety Notes

1 Take care when handling material and moulds that have just been removed from the oven.
2 With a large mould, quench it adequately so the residual heat in the centre is released.

Some Observations

Select a sample casting and check the following points:

1 What is the finish like?
2 Is there any shrinkage? Compare the component with the mould.
3 Are there any air bubbles or voids?
4 How has small detail come out?
5 Saw a casting in half and look at its cross-section.

Some Applications The following applications have been mentioned already and from these ideas the scope can be broadened, depending on the demands and facilities available for mould making:

1 Gear wheels
2 Draughts
3 Push-on feet for stools or tables

Notes on Mould Making Fig 22 shows a simple mould for polythene powder casting of a push-on foot for legs of tables or stools.

A This is a pin to produce a hole in the casting for the table or stool leg. This pin also acts as an ejector pin for the casting.
B This is a recess machined out to allow the pin to be pushed in easily.
C This taper has been left from the twist drill used in the drilling out prior to boring and could be looked upon as an aspect of design.
D This shows how the sides have been bored out to a good tapering shape, thus allowing easy ejection of the component.
E This shows the cavity that is charged with polyethylene powder

The mould surfaces should be polished out to a high finish and the pin 'A' should have the very slightest hint of a taper on it to facilitate it being withdrawn from the component.

The diameter of pin 'A' should be fractionally less than the diameter of the leg onto which the foot is to be fitted; this will mean that the foot will be a tight fit on the leg.

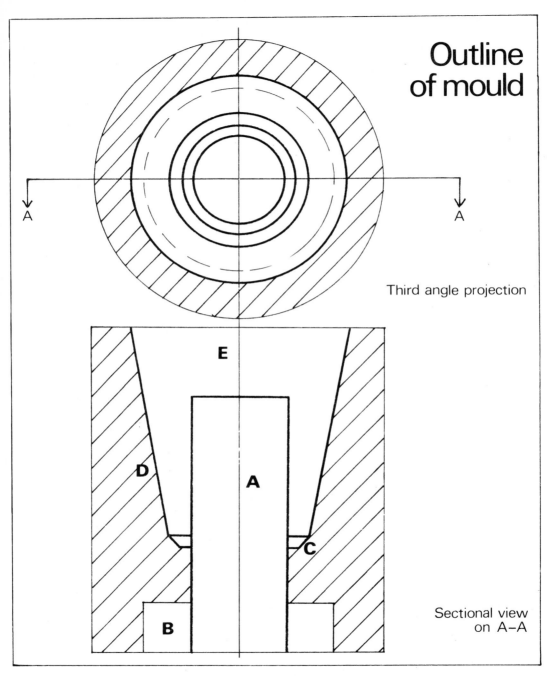

Outline of mould

Third angle projection

E

D

A

C

B

Sectional view on A–A

Fig 22

4 Plastisols

Plastisols are dispersions, emulsions or solutions of polyvinyl chloride resins in plasticizers. The plasticizer is a solvent which makes the material more flexible in its application and performance. When the resin and plasticizer combination is heated, the amount of plasticizer is reduced and the resin gels.

Polyvinyl chloride in its liquid or paste form can be used in a variety of applications and when it is heated it becomes a flexible rubber-like material. This state can be varied from very supple to quite rigid. A plastisol's properties are many and they make it a versatile material: the electrical insulation properties are good; a coating of PVC presents a barrier to corrosion; the 'feel' of articles coated with the material is pleasant; colours and finishes—matt or glossy—make it an attractive material to use; PVC gives to an article a tough and durable surface; it is waterproof; it is odourless and it does not tend to discolour.

Applications

Plastisols are used in the following work:

Hot-dip coating	Slush moulding
Hot-dip moulding	Mould making
Rotational casting	Spray coating

Hot-dip coating: This is essentially the coating of articles by heating them, dipping in a bath of plastisol, allowing the excess to drain off and finally re-heating them to gel and cure the coating. Some typical applications are the coating of dish drainers, dish-washer machine plate and cutlery holders, coating imitation wrought-iron products.

Hot-dip moulding: This employs similar principles to the hot-dip coating process. A hollow former is pre-heated and dipped in the plastisol. Re-heating takes place to gel and cure the coating and then the moulding is stripped off the former. Such articles as cycle handlebar grips and gaiters to protect steering joints on motor cars are moulded in this way.

Plate 13   *Rotational casting—two dolls*
Courtesy of ICI Plastics Division and Favourite Toys Ltd

Rotational casting: This involves moulds being rotated to form components that are hollow and flexible. These moulds are contained within an oven and are rotated in both the horizontal and vertical planes simultaneously. This action forces the plastisol into every minute detail of the mould.

58

Some typical applications are dolls, baby toys, footballs and play balls.

Slush moulding: This process has been largely superseded by rotational casting, but it is still used for workpieces where quantity is not the paramount consideration. A mould is pre-heated and then filled with plastisol; after a short while the paste is poured out, leaving a gel coating on the mould surface. Re-heating then takes place to cure the moulding fully.

Mould making: This involves the production of a mould made of the plastisol material itself. The paste is simply poured around the pattern from which the mould is made, leaving one face of it uncovered so it may be extracted later on in the process; heat is applied to gel and cure the material; finally, the pattern is removed, so leaving its own impression. Castings can then be made in this mould.

Spray coating: This process is used on large areas which require coating and where dipping would be impossible. Where drainage of excess plastisol after dipping would

Plate 14   *Dolls made in school in plastisol moulds*

present problems, spraying would be used rather than hot-dip coating. Intricate mouldings can also be produced by this technique. The process is much the same as for the hot-dip coating except that a vessel or bath is not needed to contain the plastisol but spraying equipment is used instead.

Most, if not all, of these processes can be attempted at a school level, depending on the equipment and facilities that are available.

Mould Making, Hot-Dip Coating and Slush Moulding are described here in closer detail.

Mould Making

Material: PVC paste.

Equipment: Tobacco, polish or any shallow tins, patterns —for this example a gear wheel has been chosen—an oven.

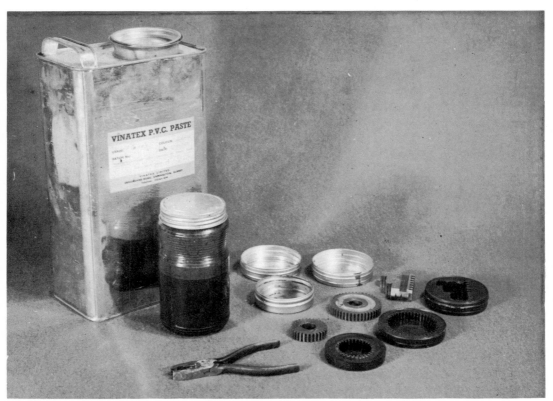

Plate 15   *Equipment for mould making*

Procedure (see fig 23)

1 Select a gear wheel and a tin to fit it—a gap, 8 mm approximately, should be left all round between the gear wheel and tin.
2 Pour some PVC paste into the tin to a depth of about 4 mm.
3 Place the tin in the oven, set to operate between 160° and 180°C.
4 As soon as this paste has changed from the liquid to the gel stage, remove the tin from the oven.
5 Place the gear wheel in the tin on this gelled surface.
6 Pour in more PVC paste around the gear wheel up to the top surface—*do not overfill and cover the gear wheel*.
7 Return the tin to the oven and cure fully. **Note**: A completely cured PVC mould should be tough and durable. Weak or flaky texture denotes under-curing and can be remedied by additional heating. Over-curing is denoted by a darkening of the material, blistering and the emission of acrid fumes.
8 Remove the tin from the oven and allow to cool for a short time in the air.
9 Quench and remove the gear wheel from the mould and then the mould from the tin.
10 This mould can now be used to cast gear wheels, using polyester resin. A selection of wheels is shown together with a practical application of them.

Safety Notes

1 Respect a hot oven and use it with care and common sense.
2 Take care when handling trays or materials that have just been removed from the oven.
3 Avoid the over-curing of PVC as the fumes can be harmful.

Some Observations

1 Inspect the mould for quality of definition.
2 Has the floor of the mould adhered well to the sides?
3 Note the flexibility of the material—moulds with undercut can be produced quite easily.

Some Applications With the production of a mould, a considerable number of components can be produced:

1 Sets of drawing-office models
2 Parts for models
3 Sectioned components, e.g. half a hexagon-headed bolt

61

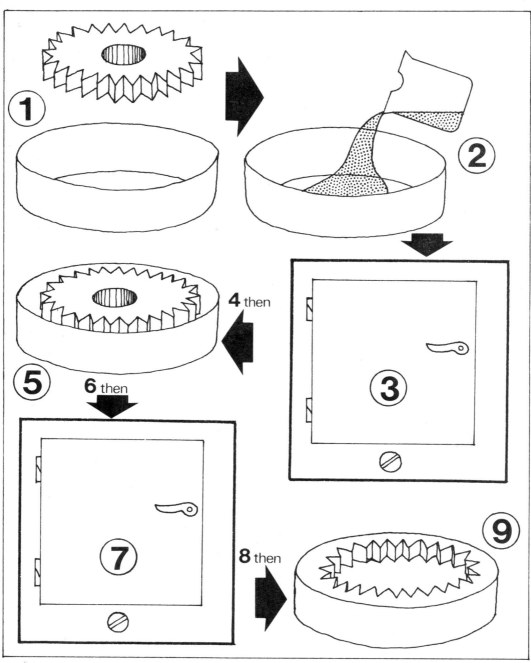

Fig 23

GEAR TRAIN DEMONSTRATION MODEL

Plate 16　　　　　　　　*An application of resin-cast gear wheels*

Hot-dip Coating　　PVC paste can be applied to a metal surface like a coat of
paint and will afford protection against corrosion, give the
article a better feel and impart a decorative, colourful finish.
PVC does not adhere to metal, but where the coating covers
the entire surface, as in the coating of wire work, the bonding
of the material to the substrate is unnecessary. If, however,
only part of the component is to be coated, a metal adhesive
should be applied to effect good bonding of PVC to the
metal surface.

Material: PVC paste, metal adhesive (if required).

Equipment: A paste tank constructed of steel, aluminium or
stainless steel (a glass container will suffice for small work),
an oven, wire, pliers, de-greasing solution.

63

Procedure (see fig 24)

1 Thoroughly clean the surface to be coated and de-grease with 1 : 1 : 1-trichloroethane.
2 Attach fine wire to the article for suspension in the oven, or devise jigging if only part of the article is to be coated.
3 Heat the component to between 170° and 200°C.
4 Remove from the oven and dip immediately—the paste temperature should not be below 17°C.
5 Extract the component at such a rate as to render an even coating and avoid run marks, layering and drips.
6 Replace in the oven and completely cure the coating. An average coating on wire 3·2 mm diameter needs 6–8 minutes to cure.
7 Remove the coated article from the oven and allow to cool in air for a short while.
8 Quench and inspect the coating.
9 Finally, cut off the suspension wire at the surface of the coating or, if the component is only part-coated, trim off coating to a clean line.

Safety Notes: These are as for Mould Making.

Some Observations

1 Using a spare component or a test-piece which has been coated, test the coating for hardness—indentation with the fingernail.
2 Test the abrasion resistance of the surface.
3 Has the coating any smell?
4 Will it peel off very easily—compare surfaces that have no metal adhesive beneath them with those treated with adhesive.
5 How does the coating react to water?

Some Applications Steel is an obvious material for treatment with PVC paste as rust and corrosion will be prevented. The feel of the coating and the ease of application also recommend its use.

1 Coating of wire three-dimensional models
2 Workshop tool finishing
3 Alternative finish for metalwork jobs

Slush Moulding This is the most specialized of the three activities using PVC plastisol material because it does require a metal mould in which the casting is made. Slush moulding gives a true

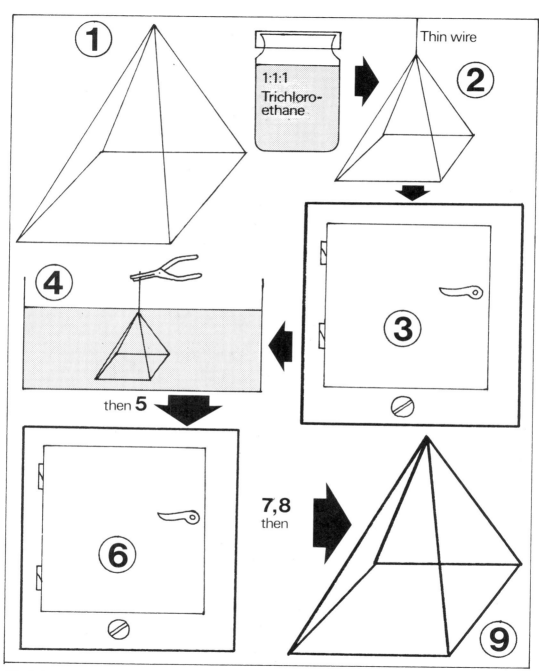

1:1:1 Trichloro-ethane

Thin wire

① ② ③ ④ then **5** **6** **7,8** then ⑨

Fig 24

65

exterior finish whereas, for a true interior finish, dip moulding should be employed. The moulds used for this work need to have only very thin walls, e.g. 4 mm. They are usually cast in aluminium alloy. Surfaces should be polished and smooth, for the PVC paste will flow into every mould feature and poor mould finish will be reflected on the finished component. A two-piece mould of the head of a doll is shown in plate 17. The production of similar moulds will require the assistance of the metalwork department. Moulds can also be produced by turning on a centre lathe or by metal spinning techniques and this could make interesting project work linked with the plastics side.

Material: PVC paste.

Equipment: A mould, asbestos gloves, pliers, an oven.

Procedure (see fig 25)
1 Heat the slush mould to about 170–185°C. For this, a simple aluminium tube with one end closed has been selected.
2 Pour in a quantity of paste and leave this in the mould for 1–1½ minutes. A wall thickness of approximately 4 mm will be obtained with this time/temperature combination.
3 Pour out the excess paste into the container for re-use later on.
4 Return the mould to the oven and cure fully for approximately 7–10 minutes.
5 Remove from the oven and quench to about 60–80°C.
6 Extract the component from the mould. *Note:* It is advisable to remove slush-moulded components from the mould while they are still warm as the increased flexibility can assist removal. Hence the quenching to about 60–80°C is important.

Safety Notes: These are as for Mould Making.

Some Observations
Check the moulded component for the following:
1 Surface finish
2 Under- or over-curing
3 Uniformity of wall thickness
4 Flexibility

Some Applications
1 The production of short lengths of tubing and pipe
2 Flexible containers
3 Flexible toys

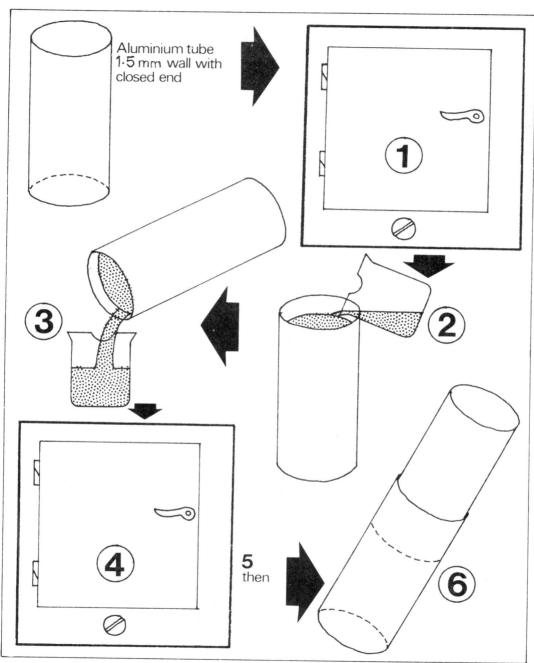

Aluminium tube 1·5 mm wall with closed end

1

2

3

4

5
then

6

Fig 25

Plate 17   *A model policeman's head made by slush moulding*

5 Acrylic Materials

One of these materials is widely known under the name of 'Perspex', which is a large manufacturer's name for acrylic sheet. These materials are thermoplastics, available in sheet, rod and tube form, also as moulding compounds. When heated they can be shaped easily; in the hard state they can be worked with conventional tools. Because acrylic sheet transmits light extremely well it is much used in illuminated

68

signs, skylights, aircraft windows, motor-car rear-lamp lenses, domestic and industrial lighting, implosion guards for the front of television screens, machine guards and so on. Models for the Technical Drawing Office are available in acrylic materials. Some school applications and uses of acrylic sheet are shown in the following pages.

Plate 18   *Illuminated signs at London Airport fabricated from 'Perspex', ICI's acrylic sheet*
Courtesy of ICI Plastics Division

69

Fabrication Methods

1 **Heating and Softening:** Acrylic sheet will become soft and pliable at temperatures around 165–175°C and can be formed, bent or twisted in this region. When it cools it retains the shape it has been given, but if it is re-heated the material will tend to return to its original form because of the elastic memory factor of acrylic sheet. So if a mistake is made in forming a shape, it may be corrected by re-heating and starting again.

Points to watch:
(a) Always heat the material uniformly.
(b) Do not try to form the material below 112–135°C, depending on the type and dimensions of the sheet.
(c) Slight shrinkage of around 2 per cent will occur.
(d) Slow cooling avoids the setting up of internal stresses and hence produces better shapes.

When acrylic sheet is to be heated, either lay it flat on a smooth surface such as tinplate, and place in the oven, or hang the sheet vertically. These methods should prevent the sheet from becoming scratched or damaged during the heating process. For the production of simple bends, a strip heater can be used giving heat along the material where it is needed and making the handling of the sheet much simpler.

2 **Bending and Forming:** As has been mentioned previously, where a bend is required in a piece of acrylic sheet, say at a 90° angle, localized heat from a strip heater should be used and this will give a crisp sharp bend; do not heat material unnecessarily! In forming, it is assumed that most of the material will be shaped in some way, so uniform heating should be employed. The mould form should be made of a material that does not change shape around the softening point of acrylic and should not draw the heat from the sheet when it is placed on it—wood, plaster, asbestos are suitable mould-form materials. When the sheet has been heated sufficiently, it is draped over the form and held in contact with it until cooling has taken place. Contact with the mould can be maintained with the help of 'G' cramps, pins or dowels, spring pressure and even pressure from a vacuum bag.

3 **Use of Hand Tools and Machining:** It must be remembered that acrylic sheet is brittle in the cold state and that heat generated in cutting operations will tend to soften it.

70

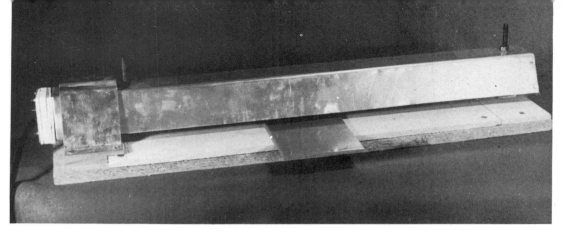

Plate 19　　　　　　　　*A simple school-made strip heater for acrylic sheet*

(a) Sawing: Conventional wood and metal saws can be used, keeping the pressure light. Jig saws can be used for curved work, slots and contouring taking care to keep the sheet pressed down well onto the saw table and employing a light feed.

Plate 20　　　　　　　　*Sawing acrylic sheet on a jig saw*

(*b*) *Filing:* Clamp the sheet firmly and keep it low in the vice. Use normal double-cut metalworking files and angle the file to the material when cross-filing.

Plate 21   *Filing acrylic sheet in a bench vice*

(*c*) *Drilling:* Twist drills should be given plenty of clearance and a point angle of 140° (approx.). Keep speeds fairly high, cuts light and feeds slow. As heat will soften this material, cool it during drilling with soluble oil, taking care that the coolant used does not attack the material. Probably most important of all, the sheet must be clamped very firmly down to prevent it riding up the flutes of the drill—if this happens cracking and even shattering of the sheet will result.

72

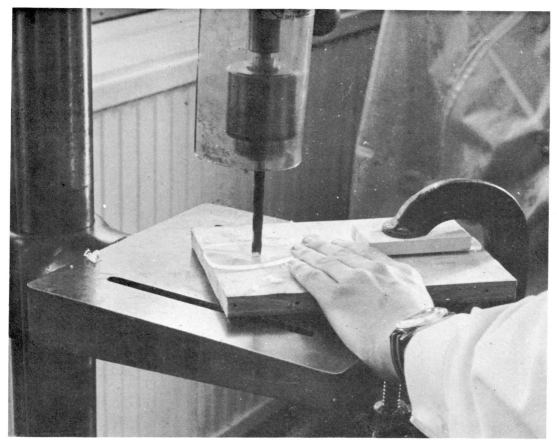

Plate 22   *Drilling acrylic sheet*

(d) **Tapping and screwing:** As for metal, using lard oil for a lubricant.

(e) **Turning and milling:** Tools should not have points but radiused tips so as to avoid surface scratching. Adequate clearance is necessary and speeds should be kept high. Again as for drilling, keep cuts light and feeds slow and use coolant if heat begins to build up.

(f) **Engraving:** The pattern to be engraved can be marked out on the paper protection on the acrylic sheet. If you are to copy a

73

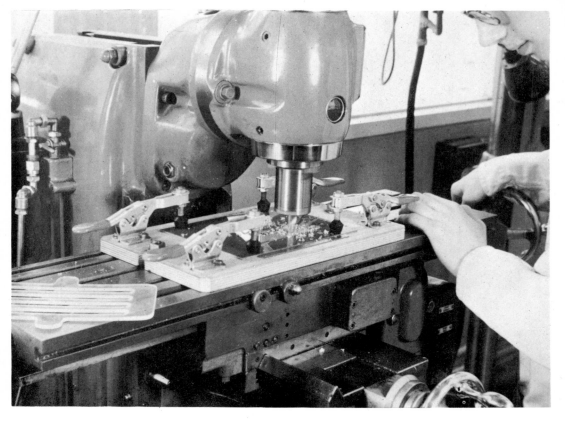

Plate 23   *Milling components for a toast rack in acrylic sheet*

pattern and a clear sheet is to be used, the sheet can be placed over the pattern and copied through.

In all the above operations, avoid scratching the sheet material ***at all costs***. Usually paper is stuck to each side of the sheet and this should be left on ***for as long as possible.***

4 Bonding or Cementing: This can be carried out by using one of the following methods:

(*a*) Using solvents
(*b*) Using specially prepared cements
(*c*) Using polyester resin, 'catalyst' and accelerator

(a) The solvents to use are those which attack the material quickly and yet evaporate quickly, allowing the bonding to take place in reasonable time. Chloroform, ethylene dichloride, trichlorethylene and glacial acetic acid all possess these properties. The surfaces to be joined should be clean and smooth and must fit together without gaps. Apply the solvent to both surfaces and apply light pressure until a bond is effected. Solvents such as chloroform and trichlorethylene *must be used in a well-ventilated area*.

(b) Specially prepared cements are marketed and these all contain a certain amount of acrylic material. They therefore generally produce a stronger joint and tend to fill spaces that may be present between the two surfaces.

(c) The use of polyester resin gives a much stronger joint. Mix resin, 'catalyst' and accelerator to the manufacturer's instructions, cover both surfaces to be joined, then bring together under light pressure. Because the resin gels quite quickly, a swift bonding is gained and a high degree of strength results.

5 Finishing: Acrylics can be sanded and polished, but in each process special care must be taken to avoid overheating. All thermoplastic materials will react unfavourably to this situation. Excessive speeds, hard buffing wheels and too much pressure should all be avoided. Where edges, tool marks and scratches are to be removed, a rotating abrasive disc or papers used in the hand can be employed judiciously so as to avoid overheating; a 254-mm diameter sanding disc should revolve at about 3000 rev/min. After this, mechanical buffing can be used, employing soft calico mops lubricated with mild abrasive such as rouge or tallow. This will bring acrylic to a high finish, but during all buffing work keep the material on the move to avoid the build up of heat.

A good point to observe always is the avoidance of damage to the surfaces of acrylic sheet. In this way finishing is reduced to an absolute minimum.

Some Observations
Select an odd off-cut of acrylic sheet and attempt the following tests:
1 Check the quality of the surface.
2 Rub it on a bench top and see how it scratches.
3 How flexible is it?

Plate 24   *Buffing a panel for edge lighting in acrylic sheet*

4 Has it a particular smell, especially when being sawn or filed?
5 How does it burn? Hold a small piece in a Bunsen flame.
6 Try the elastic memory test: heat a piece of acrylic, bend it
  and allow to cool; re-heat and observe how the piece returns
  to its original shape.

Some Applications   As can be seen from the foregoing section, the applications of acrylic sheet are many and a variety of uses will be found for this colourful, attractive and versatile plastics material.

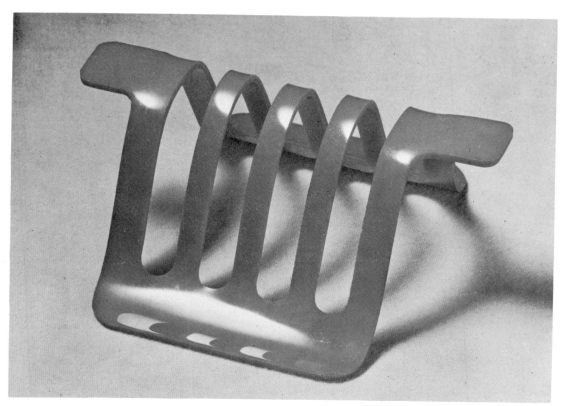

Plate 25   *School toast rack*

Section IV

FURTHER WORK WITH APPARATUS

1 Dip or Plastics Coating

Introduction

Dip or Plastics Coating is a process that today is as important as the enamelling or plating of metal surfaces and it is therefore one of the major methods of finishing metal.

Coating a heated metal in a powdered plastics which is subsequently fused gives to the metal a serviceable decorative finish and a good degree of protection against rust and corrosion.

The thermoplastics usually employed in this process are:
1 Polyethylene
2 Polyvinyl chloride (PVC)
3 Cellulose acetate butyrate
4 Nylon

Polyethylene is easy to apply, has a wide range of domestic applications, has a pleasant feel and finish and is durable and resilient enough to withstand fairly rough treatment.

Polyvinyl chloride (PVC) presents a much harder surface, it is attractive and has a pleasant feel, but it is a more expensive material than polyethylene.

Cellulose acetate butyrate is particularly useful where a degree of chemical resistance is required on the surface of the coated metal together with a resistance to abrasion.

Nylon gives a very hard durable surface, is pleasant to the touch but requires higher temperatures than polyethylene in the coating process.

Typical Applications

Polyethylene: Dish drainers, soap containers, steering wheels, trolleys and baskets for supermarkets, refrigerator shelving, general wire work, milk crates, lighting shields.

Polyvinyl chloride (PVC): Tool handles, electrical fittings.

Cellulose acetate butyrate: Door handles, hand rails, valve handwheels, coat hooks, steering wheels for motor vehicles.

Nylon: Limb appliances for disabled people.

Plate 26   *Plastics coating—dish drainer*

Historical Background to Plastics Coating

The discovery of polyethylene in the 1930s led to its wide use in the war of 1939–45 and it was the war that brought polyethylene as a plastics material to the fore. As early as 1940 engineers were looking for applications of polyethylene other than as an effective part of the radar defences of Great Britain. They duly succeeded in converting it to a powder form and in the year after the war ended, 1946, polyethylene was made commercially available in powder form.

Development work ensued and the dipping of heated metal into the powder produced the first plastics-coated articles. This process was crude and, although articles emerged coated with a covering of plastics, thickness and uniformity of coating could not be controlled or guaranteed. There was a pressing need, therefore, for the development of an improved production process. Work had been going on for some time on fluidization, that is, a technique to make a solid in particle form behave like a liquid. A German company adapted powder fluidization to plastics-coating techniques in the early 1950's and plastics or dip coating became a commercially viable process.

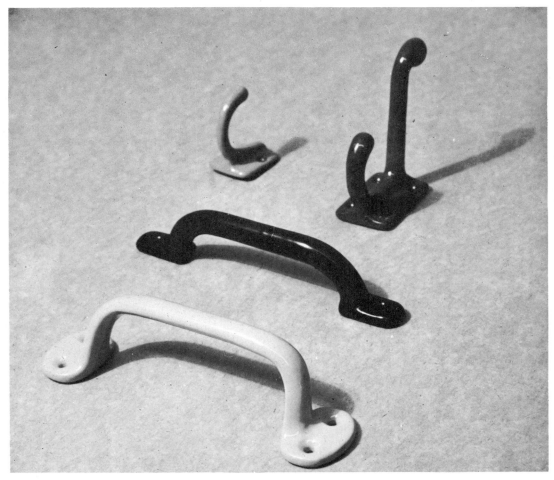

Plate 27   *Plastics coating—hooks and handles*
Courtesy of Telcon Plastics Ltd

Fluidization   There is an experiment in physics where a container of sand has placed on its surface a table-tennis ball and an iron weight. They both rest on top of the sand as if they are floating. Now

82

air is gently blown through the sand, and whilst the table-tennis ball remains floating, the iron weight sinks just as if the two objects were now in a liquid or fluid. This experiment demonstrates differing densities and the fact that particles become less compacted when air is blown through them. It also demonstrates fluidization, which is so important in dip coating.

It was stated earlier that dipping articles in a quantity of powder produced haphazard results, where coatings were uneven. Parts of intricate components might receive no coating of plastics powder to speak of. When the polyethylene powder is fluidized it is as if the article to be plastics coated is dipped into a fluid; it becomes evenly coated as it would become evenly wetted in a liquid, so that even the most complicated shapes and confined areas are treated.

Fluidization of plastics powder is achieved by the separation of the powder from the air by means of a porous medium—a ceramic tile or heavy canvas—with extremely small pores which allow air to pass through in an even mass. This air then permeates through the mass of powder thus separating the individual particles and giving them more room to move about. This imparts to the powder a fluid-like character.

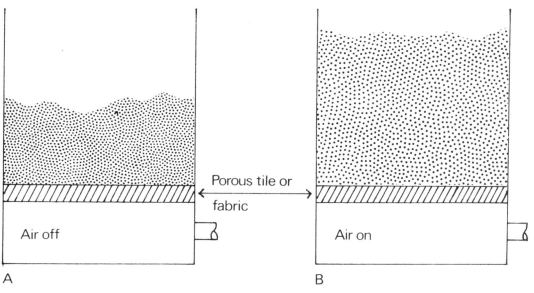

Porous tile or

fabric

Air off

Air on

A

B

Fig 26

**Dip Coating
in School**

For practical purposes in school, polyethylene is the easiest
and most versatile material to use in dip coating; it requires
reasonably low temperatures for both pre-heating and fusing,
and this alone makes it pre-eminently suitable and safe for
pupils to use.

Material: Low-density polyethylene powder.

Equipment: Oven, pump/compressor unit, fluidizer unit,
de-greasing solution, pliers, asbestos gloves.

Plate 28   *Dip-coating equipment*

Procedure
1 Articles to be coated need to be free of any loose surface
deposits—rust, scale, etc. Also they should be de-greased

thoroughly so that the plastics coating adheres satisfactorily.
1 : 1 : 1-Trichloroethane is suitable for this, and it should be
used in a well-ventilated area.

2 Articles are placed in a suitable sized oven and heated. The
temperature will vary with the mass of the article, e.g. thin
wire will need to be heated to a greater extent than thicker
material. An average temperature works out at approximately
180°C.

Plate 29   *De-greasing the article to be dip coated*

3 When the correct temperature has been reached and maintained, the articles can be quickly transferred one by one to the fluidizer and dipped. The diagrams show how a fluidizer operates: Fig 26 (A) shows the powder in an unfluidized state and (B) 'fluidized' with the air being forced through it, thus making it behave like a fluid. This fluidized powder will coat articles evenly and completely.

Plate 30   Dipping in the fluidizer

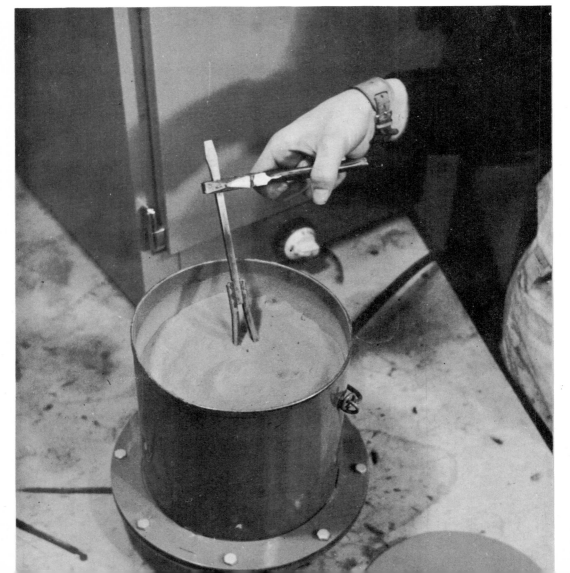

4 After dipping for a few seconds, surplus powder is tapped off and the articles can be returned to the oven. This will cause the powder which has adhered to the surface to fuse or melt. On larger articles fusion may take place with the residual heat in the article coming through to the surface and fusing the powder.

5 Cooling of the articles can be carried out either by quenching or by allowing slow cooling in the air. The latter technique does produce a flatter and better-looking finish on the component. Care should be taken to see that the plastics-coated surface is not touched before complete cooling has taken place or damage will result.

Plate 31　　　　　　　*Coated component*

Safety Notes

1 Use the de-greasing solutions in a well-ventilated atmosphere and take note of any health hazards they present.
2 No other special precautions are called for other than care in handling heated articles; avoid touching hot plastics material.

Notes on Jigs and Jigging

For each and every workpiece or component some device will be needed to support it in the heating phases as well as in handling and dipping. Some articles can simply be gripped with pliers, but where this presents problems a common method is to attach a very thin wire to the article, the wire being sufficiently strong to take the weight. This wire can be used to suspend and handle the article and finally can be cut

Plate 32   *Jigs used in the heating of dip-coated components*

off at the surface of the coating. However, on many school workshop jobs or components holes either plain or tapped can be used to handle the article and jigging can be done at such points. If the whole of the article is not to be coated the part that is to be left uncoated can be used to support and grip during coating. Some components are shown in plate 32, together with the method of jigging employed.

Repairs to and Removal of Plastics Coatings
Sooner or later someone will knock or drop a plastics dipped article before the coating has cooled, thus damaging it. Small repairs and patching up can be done with a hot-air blast from a domestic hair dryer or by using a naked flame, although this method can burn the coating if the heating is too fierce or too lengthy. If the coating is a complete failure it is a simple job to burn off the plastics material with a brazing torch flame. Position a tray under the article so as to catch any molten material as burning off proceeds, then wire brush the surface and repeat the dip-coating procedure.

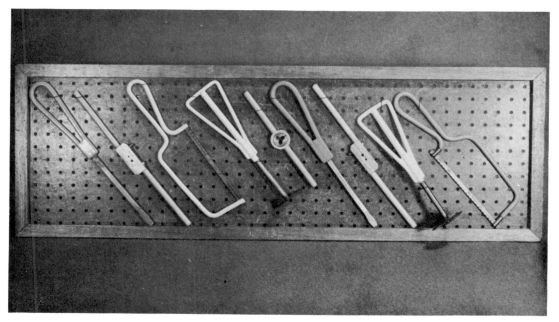

Plate 33   *A selection of workshop tools dip coated in school*

2 Centrifugal Casting

Introduction

All children have ridden on roundabouts, 'Chariot Racers', and perhaps even the 'Rotor' at a fairground. They may also have seen 'Wall of Death' riders there. All these events give fun and enjoyment, but they also demonstrate one very interesting and important scientific principle—the fact that when any quantity of matter revolves, particles of that quantity of matter try to escape and fly off outwards in all directions. Water spurting off a bicycle tyre in the rain demonstrates just this point. The force that is imparted to the quantity of matter that is revolving is known as *centrifugal force* and this force may be employed to cast shapes from liquid plastics. The plastics material most suitable for centrifugal casting is a thermosetting material—polyester resin—as used in the resin cast work described in an earlier section.

The Apparatus

A metal tube is used as a mould for the polyester resin. This tube can be any diameter and any reasonable length—usually diameters from 76 mm to 127 mm are used and lengths up to 456 mm can be handled easily. The ends of the tube should be machined off square so that end caps to be made subsequently fit accurately and are leakproof. The tube can be made of steel, copper or brass and should have a wall thickness of about 1·5 mm. The bore of the tube should be polished out to a smooth high finish—such a finish will impart to the polyester resin an equally smooth glossy surface. The resin will, of course, take on each and every feature of the bore of the tube.

The end caps can be made up from solid material—bright mild steel—but this requires a considerable amount of machining for each end cap. A better approach is to cast them in aluminium alloy where any available scrap material may be used. For this process a pattern may be made up in wood, plastics rod or expanded polystyrene. A third way of making up end caps is to machine them up from plastics rod. This method is quick and easily carried out and has an added advantage because the material releases itself automatically from polyester resin (see the sections on Fabrication from

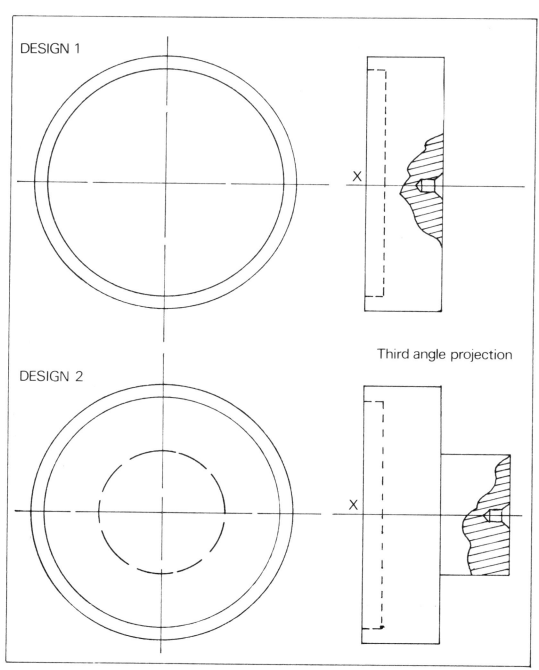

DESIGN 1

DESIGN 2

X

X

Third angle projection

Fig 27

91

Expanded Polystyrene and Machining of Plastics). A casting will require less machining and, of course, only the surfaces that are important need to be machined. Two possible designs for end caps are shown in fig 27 and, depending upon the material available and the capacity of the lathe chuck to be used in the centrifugal casting, a choice can be made.

In the machining of the end caps it is essential to achieve a good fit between the outside diameter of the tube and the recess marked 'X' on the drawings, and careful checking is necessary between these two parts during the machining operation. A 'good fit' means that a push-on fit between the tube and the recess should be aimed at, bearing in mind that with the continual use a slight loosening up will occur. Also, if the two fit together too loosely initially, they will allow the resin to leak out during casting and as time progresses the fit will become looser and the leaking more pronounced.

A centre lathe with a three-jaw chuck with universal or reverse jaws and a revolving centre complete the equipment needed.

Materials: Polyester resin (lay-up grade), accelerator, 'catalyst', pigment, fillers as required.

Equipment: Large-diameter tube with polished bore, wax release agent, two end caps, mixing containers, stirrers, paper towels, acetone.

Procedure
1 Coat the smooth polished bore of the tube with the wax release agent—Simoniz Car Wax has been found to be quite suitable. Also coat with wax the recesses of the end caps.
2 Mix the resin and accelerator to the manufacturer's specification; add the pigment and fillers if required (with fillers less resin is necessary). The amount of resin for any casting can be calculated according to the wall thickness wanted, final length of the tube and the relevant diameters.
3 Cap the end of the cylinder that is to be fitted into the chuck.
4 Add the 'catalyst' to the resin according to the manufacturer's specification and stir well.
5 Pour the resin into the tube and fit on the other end cap.
6 Set up in the centre lathe to resolve at approximately 300 rev/min. Place a small quantity of resin scraped from the mixing container on one side so that the progress of the

centrifugal casting can be checked—when the sample gels
and hardens off a little it may be assumed that the casting
is completed.
7 While the casting gets under way, remember to clean out all
containers and stirrers that have been used.
8 A limited amount of external heat can be applied to speed up
the curing process, e.g. frictional heat or electrical radiant
heat, but do not overdo this as too rapid curing can spoil the
casting.
9 After some 15–20 minutes the resin should have cured,
although times can vary quite considerably, depending on
amounts of 'catalyst' and accelerator used, and the ambient
temperature.

Plate 34   *Centrifugal casting*

10 Remove the tube from the lathe and take off the end caps. The casting should slide out. If it will not, allow the tube to cool for a while so that the casting can shrink away from the bore. If the casting still refuses to come out, place the tube and casting in an oven set to operate at approximately 150°C, allow to heat up and then remove from the oven. Stand the tube and casting in a sink and fill the centre of the casting with water, thus causing the casting to shrink from the bore of the tube.

11 One or two ends can now be cast on the cylinder, incorporating in one of the ends a threaded piece for a bulb holder if the casting is to be a table lamp. Simply mix more resin, 'catalyst', dye-stuff and accelerator, stand the casting on a surface coated with wax release agent and pour in.

Safety Notes

1 The 'catalyst' and accelerator *must* be stored and always kept *well apart* as they form an explosive mixture.

2 All traces of resin, 'catalyst' and accelerator should be removed from the skin promptly. Rubber or disposable polyethylene gloves should be worn by anyone prone to skin complaints.

3 Resin castings are brittle—handle them with care!

Some Applications   Centrifugal castings can be put to a variety of uses and practical applications. Table lamps, with or without weighted bases, small trinket boxes and lamp shades can be made.

Important Note: Some resins are sold in a pre-activated condition—that is, accelerator, activator or promoter has been added and all that is needed to initiate polymerization is the catalyst.

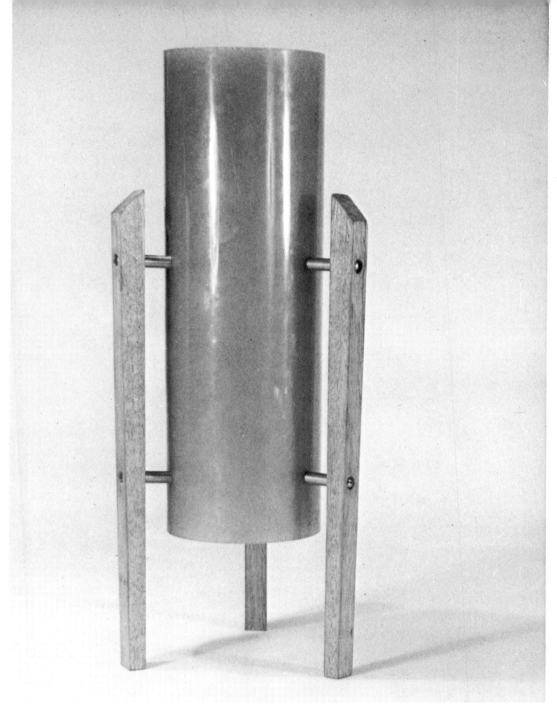

Plate 35   *A table lamp made by centrifugal casting in school*

3 Compression Moulding

Introduction

This is a process whereby a plastics material is shaped under the influence of pressure and heat—heat because all plastics need heat to enable them to undergo a physical change and pressure to force the material tightly into the mould form. This technique of moulding can be traced back a great many years to such operations as moulding clay in the hands and placing it in the hot sun to bake and harden, or forcing clay into rectangular moulds and baking it in the sun to make building bricks—both a sort of compression moulding.

Both thermosetting and thermoplastic materials can be compression moulded, but the process tends to apply more to the thermosetting materials.

The uncured material may be in a powder or granule form and may be mixed with a filler to give added strength as well as economy on material; such fillers as wood flour, powdered metals, mica and clay are used. The common thermosetting plastics materials used are phenol-formaldehyde, urea-formaldehyde and melamine formaldehyde resins.

In this form of moulding it is important that a carefully measured quantity of material is placed in the mould before each moulding cycle commences, for too much or too little material will result in a spoilt component. A system has been developed for industry where tableting or preforming of the material is carried out so that an already partly compressed and partly shaped mass of material is inserted into the mould form, to be finally shaped and cured in the compression moulding cycle proper. This introduces another process, so adding to costs, but it is necessary for efficient and good-quality moulding.

A modification to compression moulding is transfer moulding, where the moulding material is pre-heated in a cavity next to the mould form and when the material becomes plastic and starts to flow it is forced under pressure into the heated mould proper. This is a very similar technique to injection moulding of thermoplastic material, which will be described in a later section.

96

Plate 36

An electricity-meter compression moulded in Bakelite
Courtesy of Bakelite Xylonite Ltd

The Apparatus

The apparatus needed for compression moulding can be quite simple as long as it fulfils two specific requirements: (*a*) It can exert pressure up to 5000 lb/in^2. (*b*) It has built into it, or is capable of having attached to it, an electrical heater unit,

sufficiently large and powerful to heat the complete mould.
As will be seen from the Outline for a Compression Moulding
Machine, fig 28, there are some six main parts.

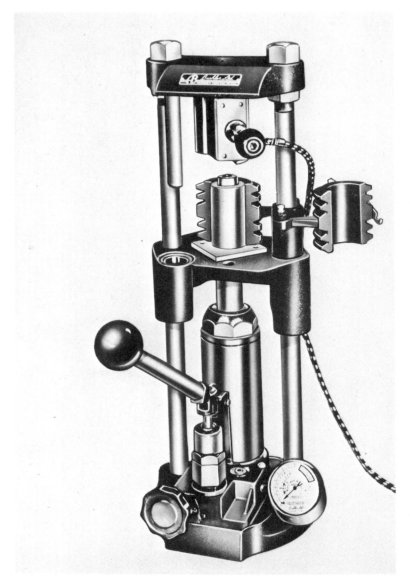

Plate 37   *The A.B. Specimen Mount Press*
Courtesy of Buehler Ltd

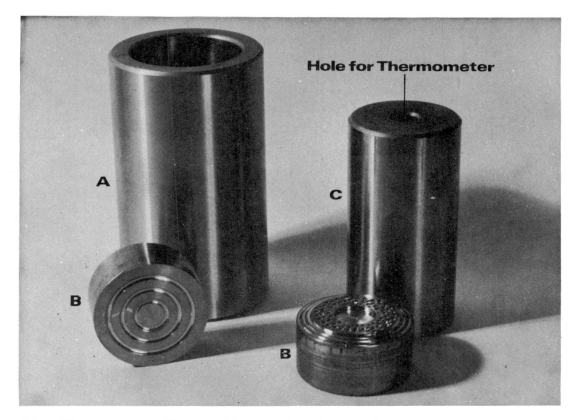

Hole for Thermometer

A

C

B

B

Plate 38   *Two moulds, for a draught and a button*

The Mould

The easiest moulds to produce are circular ones and quite a good deal of scope lies in the circular mould form alone. Such moulds can be made in the school workshop and so the process of compression moulding can be followed through from start to finish. Two moulds are shown in detail in plate 38—one for a draught, the other for a button. A mould comprises three parts:

A Mould cylinder
B Mould base
C Mould ram

Parts B and C must be a very good sliding fit in part A, and all surfaces of the mould parts should be polished. Lack of care in fitting will result in the mould leaking when under its normal

99

working pressure. Pressures during compression moulding vary between 2000 and 5000 lb/in^2 and temperatures in the region of 150–200°C. Times taken, for the material to cure fully, range from seconds to perhaps 15 minutes.

Materials: Phenol-formaldehyde resin or urea-formaldehyde resin.

Equipment: Compression moulding machine, mould release agent, asbestos gloves, stop clock or watch.

Procedure
1 Check that the mould parts are clean and ensure that there are no remains of the last moulding.
2 Apply the mould release agent—often a silicone-based oil spray.
3 Fit the mould base into the mould cylinder.
4 Charge the cylinder with a measured quantity of powder—this quantity can be found by experiment—a useful and interesting piece of research for pupils.
5 Insert the mould ram—use care in inserting it into the cylinder and **do not** force it if it sticks.
6 Place the assembled mould on the sliding platen and line up the ram on the upper platen with the mould cylinder—this is most important as the ram on the upper platen must enter the cylinder cleanly so that it bears on the mould ram.
7 Surround by the heater unit and switch on.
8 Apply pressure to approximately 4000 lb/in^2 and hold this. As the powder or granules settle down air escapes and the pressure will drop, so watch for this and maintain pressure.
9 A sharp drop in pressure will indicate that the material is flowing. Maintain the pressure when this happens. Keep the pressure applied until the whole mould has reached the temperature applicable to the material being used, e.g. 150°C for phenol-formaldehyde. This allows the resin to cure throughout.
10 Switch off the heater and release the pressure.
11 Handling the mould with asbestos gloves, eject the mould base, component and mould ram. An ejecting mechanism can be built onto the machine.
12 Allow the component to cool and prepare the mould for the next shot.

Note: The curing time is very important and both under-curing and over-curing will result in a bad component.

100

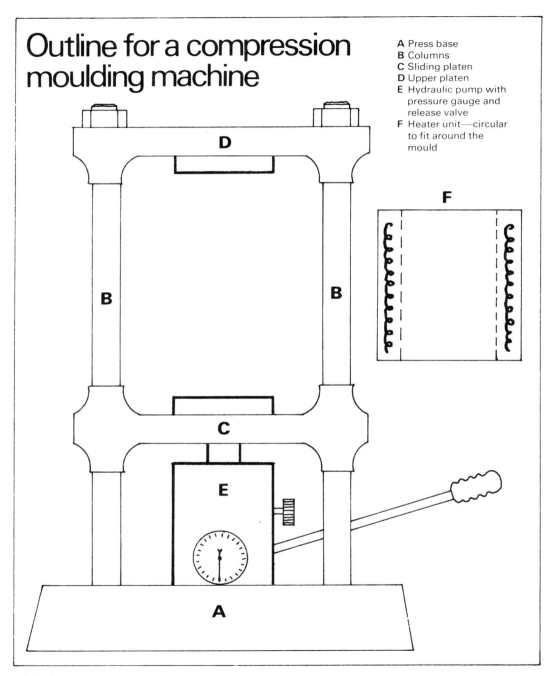

Outline for a compression moulding machine

A Press base
B Columns
C Sliding platen
D Upper platen
E Hydraulic pump with pressure gauge and release valve
F Heater unit—circular to fit around the mould

Fig 28

Safety Notes

1 The machine must be constructed to withstand the stresses and strains of pressures of up to 5000 lb/in^2.
2 No special safety precautions are called for other than care in handling a mould and component that may be at temperatures in excess of 150°C.

Plate 39   *A selection of commonly known articles made by compression moulding*

4 Blow Moulding

Introduction

It is fascinating to watch a highly skilled glass blower at his work, forming intricate and precise shapes with air pressure supplied by his mouth. He is, in fact, blow moulding glass, and although he may use other tools to achieve the desired design, he essentially moulds the article by blowing. With

Plate 40

Blow moulding—industrial machinery for the manufacture of polyethylene bottles
Courtesy of ICI Plastics Division

plastics, blow moulding involves forming a shape by blowing the material which has been heated into a mould of that shape.

Most thermoplastics materials can be used for blow-moulding work—the more common ones are:
Polyethylene
Polystyrene
Polyvinyl chloride (PVC)
Polypropylene
Acrylic sheet
Acrylonitrile butadiene styrene (ABS)

On the commercial side, blow moulding is made up of two linked processes:
1 The production of a tube or parison by an extrusion or injection moulding process.
2 Blowing the tube or parison to make it conform to the shape of the mould.

The sequence of blowing a bottle or container is shown and comprises three stages: in fig 29.1 the tube or parison is introduced into the mould form straight from the head of an extruding machine. Fig 29.2 shows that the mould form has closed and has shut off the open end of the parison. Finally in fig 29.3 the 'blowing up' of the tube is performed and the job is complete. The mould form must be split into two parts so that the component can be extracted after blowing.

The idea of blowing air into a soft, hot plastic mass is not new; at around the turn of the century—1900 or so—celluloid dolls, rattles and table-tennis balls were all produced by blow moulding. During the last war, aircraft 'blister' windows were produced by clamping a sheet of plastics material, usually acrylic sheet, heating it and blowing gently underneath it, forming it into part of a sphere. The pressure was released when the sheet had cooled sufficiently. This process could really be called 'free-space' blow moulding, for the material is not being blown into a mould form but just into space. The resulting shape will be governed by two things:
1 The shape of the mould plate perimeter
2 The amount of air blown in

'Free-space' blow moulding requires simple apparatus and is an activity which can be easily carried out in school. Very

104

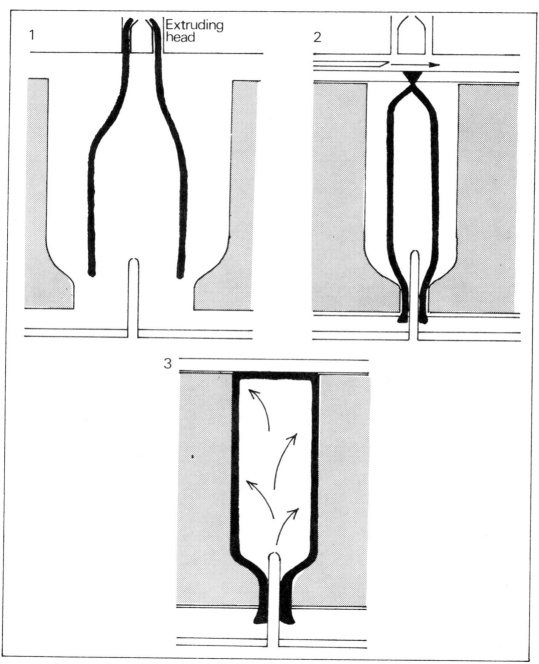

Fig 29

often it can be extremely cheap in that one can use up odd scraps of plastics sheet which would otherwise be thrown away. (Industrially this is a type of thermoforming.)

The Apparatus

A 'free-space' blow moulder comprises two main parts:
1 A base plate
2 A mould plate

These two are firmly held together with some screw or clamping system; the base plate contains the connection to the pump which supplies the compressed air in the blowing stage; the mould plate has the shape of the perimeter of the blow moulding cut in it. The two plates must fit together accurately so that no air will be lost between them during the blowing of the shape. Reference to plate 41 will give the idea of a basic 'free-space' blow moulder. This sort of apparatus can be made up in the school workshop and a range of mould plates can be made up all to fit onto the base plate, making the unit adaptable to several blow-moulding activities.

Plate 41

A 'free-space' blow moulder with selection of plates all made in school

Blow Moulding in School

Materials: ABS sheet, acrylic sheet, polypropylene, polyvinyl chloride (PVC), polystyrene and polyethylene.

106

Equipment: 'Free-space' blow moulder, asbestos gloves, oven, pliers, hand pump.

Procedure

1 Cut the sheet to the outside dimensions of the 'free-space' blow moulder.
2 Drill holes for the clamping screws. A supply of blanks can be prepared and jig-drilled ready for use.
3 Clamp the sheet firmly between the plates.
4 Place in the oven—the temperature of the oven should be commensurate with the type of material and its thickness, but an average figure for temperature is 170°C.
5 After a few minutes, remove the apparatus from the oven. The sheet will have softened and given somewhat at this point and each of the clamping screws will need an additional turn to take up the slack. Do not forget to use the asbestos gloves for this operation.
6 Continue the heating for a few more minutes and then connect the pump to the apparatus whilst it is still in the oven —this will prevent a dramatic drop in the temperature of the sheet. Give a few strokes on the pump to see if the sheet is sufficiently plastic for forming.
7 Remove from the oven if the trial pumping shows that the sheet is ready. Blow the shape to the required height and hold the pressure. Re-heat if the desired shape is not achieved.
8 Cool the apparatus—the use of a hair dryer on 'Cold' blow will be found to be quite useful.
9 Remove the moulding from between the plates and trim it to suit its application.
10 Finally, it should be noted that all this work can successfully use up odds and ends of material which would normally be regarded as scrap, hence costs of this activity can be very low indeed.

Safety Notes

No special precautions are needed other than to avoid coming into contact with hot mould plates and hot plastics material.

Some Ideas for Blow Mouldings

1 Lids for containers
2 Tea strainers
3 Egg-white separator
4 Dish made from two mouldings
5 Geometrical models

Outline for cavity blow moulding

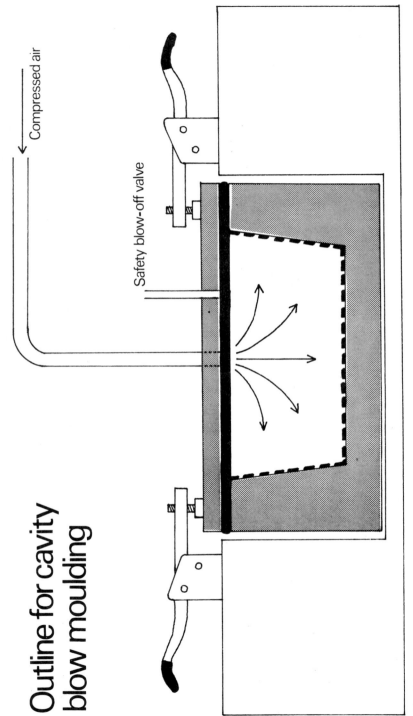

Compressed air

Safety blow-off valve

Fig 30

6 A globe of the world

7 Cups for an anemometer—a wind gauge

The reverse of 'free-space' blow moulding is the blowing of a sheet of heated plastics into a cavity or mould, and this can be carried out as easily as 'free-space' blow moulding—see fig 30.

A cavity with adequate draft is required, a plate to fit over it with a point on the plate for a connection to a compressed air line and toggle clamps to hold this plate down firmly onto the top surface of the mould with the heated thermoplastic sheet in between the plate and the mould. The sheet is heated in an oven until its softening temperature is reached, then rapidly transferred to the mould where the plate is clamped onto it and the compressed air introduced. The sheet will be blow moulded to the details of the mould cavity. The addition of a safety blow-off valve ensures that pressures used in blow moulding can be controlled.

5 Fabrication from Expanded Polystyrene

Introduction

All plastics materials have innate physical properties—some are brittle, some are flexible, some are hard and durable and so on. But these properties can be expanded—an apt word here—when additions to the material are made. A plastics material is expanded by adding an agent to make the plastics foam just like soap or detergent suds. The plastics material expands and takes on a clearly visible cellular structure. This foaming adds a new dimension to the material.

Foamed plastics are produced in several different ways and some foams will be very familiar. The chart overleaf shows some common plastics that can be foamed—note that both thermosetting and thermoplastic materials are used.

When soap suds begin to foam up, air is entrapped within each bubble, making the volume of soap or detergent appear to be much larger. So with foamed plastics, air or a gas becomes trapped in the material by the addition of some chemical agent to promote foaming. Air may be mixed mechanically

109

Base Material	Name of Foam	Applications
Polystyrene	Expanded Polystyrene	Art Work, Window Displays, Packaging, Insulation in the Home
Polyurethanes	Polyester Polyurethane Foam	Buoyancy in Boats, Art Work, Reinforcing Radomes, House Insulation
	Polyether Polyurethane Foam	Synthetic Sponges, Cushions
Polyvinyl Chloride (PVC)	Plasticized PVC Foam	Cushioning and General Upholstery Work
Polyethylene	Polyethylene Foam	Electrical Cable Insulation and Buoyancy for Sea-borne Cables

into the plastics; gas may be produced during the mixing process; a 'blowing agent' may be incorporated in the material, so that when a certain temperature is reached the agent breaks down and gas is given off.

The resulting material has several properties that the former base plastics did not possess:
1 It is lighter volume for volume—less dense.
2 It takes up much more space consequently.
3 It can be stronger.
4 It has a cellular structure which may be open or closed or a mixture of each. A structure is 'open' when the cells interconnect whereas a 'closed' structure has cells that are separate from one another.

An illustration of the change in weight to volume ratio may be obtained with expanded polystyrene. Place some styrene beads in a small tin which has a steam supply connected to it. After some 2–3 minutes of passing steam through, the beads will have expanded to many times their original size. The expanded beads will be hollow and it is worth comparing say 5 g of beads with 5 g of ones that have been expanded, noting how the volumes differ.

Expanded polystyrene foam is widely used for ceiling tiles, insulation of water tanks and general loft insulation in houses, and packaging of delicate instruments such as cameras, lenses

Plate 42   *Expanded and unexpanded beads of polystyrene*

and such like. This same material can be used in school for a variety of artistic and practical activities. Incidentally, people often comment when touching expanded polystyrene about it feeling warm. If some thought is given to this statement, it will be obvious that the material is not warm, but because it is a good insulator heat from the hand cannot be conducted away into the material and it feels warm.

Working with Expanded Polystyrene

The material can be obtained in a variety of shapes and sizes:
1 In rolls produced as a very thin sheet similar to wallpaper
2 In sheets of various lengths, widths and thicknesses
3 In blocks and cubes—the one shown (plate 43) being held by a boy measures 6 ft × 2 ft × 1 ft.

111

Plate 43   *This block of polystyrene is very light despite its size*

The shaping of this material presents some problems, for due to the open cellular nature of its structure any pressure exerted on it tends to make it crumble and collapse; the pressure of a knife blade or the cut of a saw tends to push the material aside rather than cut cleanly through it and the resulting surface is very rough and jagged. The material cannot be subsequently sanded or smoothed as this will only cause more breakdown of the surface. The only successful method is to use heat to melt the plastics foam apart—polystyrene being a thermoplastic it will soften and subsequently melt under the influence of heat. The heat can be applied to the material by way of a heated wire, kept hot by a gas flame, the wire being secured in a wooden handle. This method can be quite useful for sculptural work where perhaps several pupils are engaged on one piece or separate pieces of work. Such a tool, together with a variety of wire inserts shaped to give different carved forms on the foam, is shown in plate 44. This wire must be re-heated periodically and a better method involves the use of a wire which is heated electrically, thus giving a continuous source of heat.

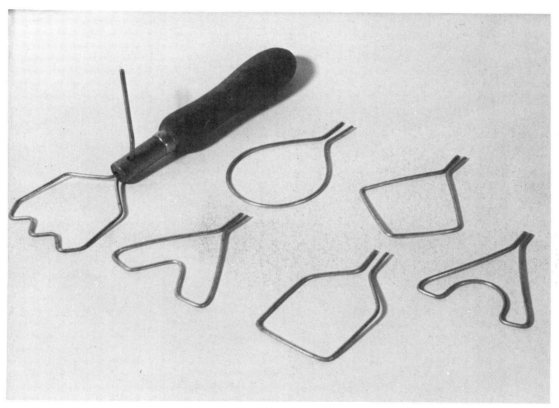

Plate 44　　　　　*A simple hand tool with a selection of wires*

The Apparatus　　　The electrically heated wire can be employed to cut foam in two ways:

1 Fitted to a machine—a hot-wire machine
2 As a portable tool—a hand-sculptor tool or saw

A hot-wire machine is basically a flat table through which the wire passes, one end of it being attached to an electrical

113

contact under the table, the other being fixed to one end of a 'U'-shaped tube which is in turn attached to the back of the table. The wire is heated by applying normal domestic voltage (200–240 volts a.c.) to a transformer built into the machine to give a low voltage and high amperage. The wire has to withstand a high temperature over long spells as well as heating and cooling when the machine is switched on and off. It must be made of nickel/chrome steel usually about 28 S.W.G. The amperage can be varied by tappings on the transformer or by operating a multi-position switch so that various thicknesses of foam can be cut. A commercially

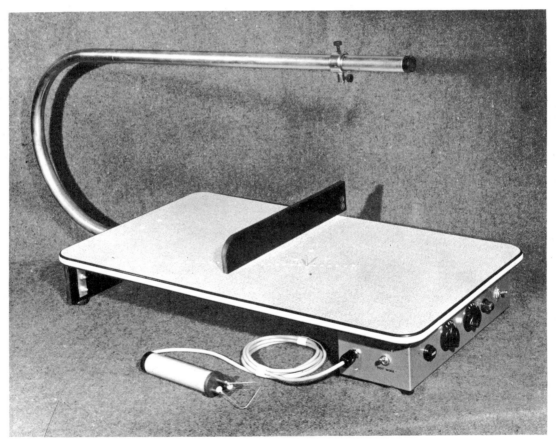

Plate 45   *A hot-wire machine*
Courtesy of Almik Displays Ltd

produced machine which incorporates all the points just mentioned is shown in plate 45. A school project to build a hot-wire machine might involve the Physics, Metalwork and Woodwork departments.

The machine in the form described allows the operator to cut the material adequately enough but essentially freehand. Some work will require a measure of precision, in cutting straight lines and angles or curves and circles. Two modifications can be incorporated into the machine for this purpose:

1 An adjustable fence, similar to the type fitted to a circular saw table. A fence will permit the cutting of parallel edges and regular shapes; it could be fitted with a protractor so that it could be set for angular cutting.

2 A circle-cutting attachment, which is a simple but very effective way of cutting curves and complete circles on the machine. A series of holes are drilled at 12 mm intervals, radiating from the hole where the wire enters the table. A pointed pin can be placed in the appropriate hole and the circle cut by revolving the material upon the pin against the hot wire.

Electrical Materials for Hot-wire Machine	Quantity	Name of Component
	1	Universal L.T. Transformer
	1 doz.	4 mm Insulated Sockets (Red)
	1 doz.	Banana Plugs (Black)
	1	Heavy-duty Toggle Switch DPST
	1	Panel Neon Red
	25 yds	40/0·0076 PVC Yellow Wire
	1 box	Instrument Case Feet, 18 mm diameter
	1	Panel Fuse Holder

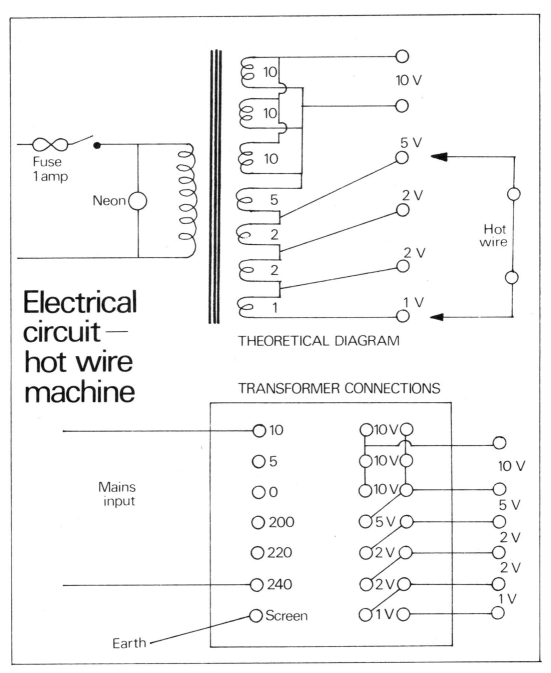

Fuse
1 amp

Neon

10
10
10
5
2
2
1

10 V
5 V
2 V
2 V
1 V

Hot
wire

Electrical circuit— hot wire machine

THEORETICAL DIAGRAM

TRANSFORMER CONNECTIONS

Mains
input

10
5
0
200
220
240
Screen

Earth

10 V
10 V
10 V
5 V
2 V
2 V
1 V

10 V
5 V
2 V
2 V
1 V

Fig 31

116

Section V

THE USE OF MORE MACHINERY

1 Vacuum Forming

Introduction

In vacuum forming, a thermoplastics sheet held firmly in a frame is heated to its shaping temperature and is then moulded by removing air between it and the mould form.

Pupils learning physics, will know that atmospheric pressure is acting on everything. At sea level this pressure can support a column of mercury 760 mm long. If a polyethylene bag is obtained and air is sucked from it a partial vacuum is created. Atmospheric pressure is acting on all surfaces of the bag— inside and outside. By reducing the atmospheric pressure inside the bag an imbalance of pressures is set up and the result is that the bag flattens out.

The vacuum forming process makes use of atmospheric pressure in such a way that, by removing air from underneath a thermoplastics sheet, it is pushed down onto the shape of the mould form.

In fig 32, 1 shows the sheet clamped over the mould form and being heated; 2 shows the sheet formed over the mould by a vacuum being created under the sheet. There are a great many articles and components produced by vacuum forming today.

The Apparatus

The machinery used in vacuum forming varies considerably, but the following basic parts are common to all equipment:

Vacuum Pump: This can be hand-operated or operated by electric motor. For school machines of a small nature the hand-operated pump is quite adequate. The pump should be the double-acting type—that is, it evacuates air on both strokes of the plunger. If a motor-operated pump is available, it should normally be connected initially to a vacuum reservoir which in turn is connected to the machine. In a school situation, the air-intake side of a brazing hearth, a paint-spray compressor or a domestic vacuum cleaner fitted up to 'suck' can all be used to supply a vacuum forming machine with the necessary vacuum. The vacuum reservoir facilitates the production of a quick and complete moulding before the sheet

118

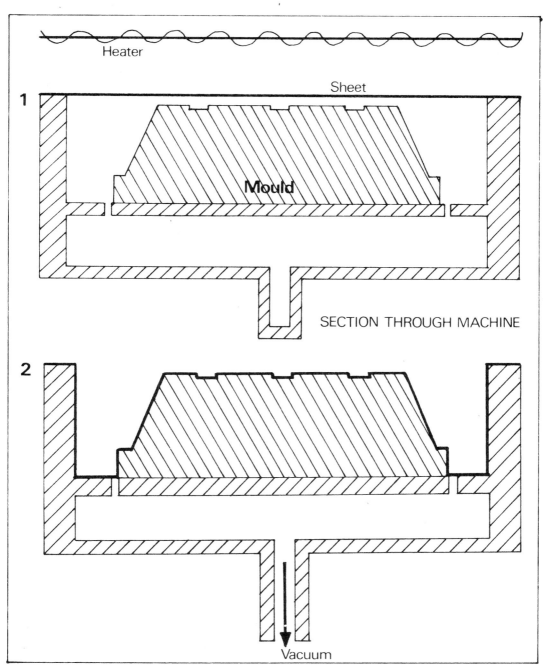

Heater

Sheet

1

Mould

SECTION THROUGH MACHINE

2

Vacuum

Fig 32

material cools below its forming temperature. The reservoir capacity should be three to four times the volume of air that will be extracted from the largest mould that will be used on the particular machine.

Heaters: Sheet-material manufacturers can generally advise on the quantity of heat that is needed to soften their particular plastics material to the shaping temperature. The amount of heat needed will depend on the thickness of the material, type and the size of the area to be heated. The type of heater commonly used is the infra-red dull emitter variety and this should be housed in polished aluminium reflectors. The heater should be so positioned as to give uniformity throughout the whole area and thickness of the sheet. The clamping frame will absorb heat and this loss should be minimized by making sure that the heater extends to cover the frame. As a general guide to the electrical rating of the heating side, allow 1–3 kw/ft^2 of material to be heated. This will cover most school needs and can effectively heat sheet material up to approximately 2 mm thickness.

The rest of the equipment comprises a table—which should be large enough for the anticipated work—and clamping frames to fit the table. The edge of the table is covered with sponge-rubber strip, as should be the clamping surfaces of the pair of clamping frames. This sponge-rubber strip will ensure that there will be no loss of vacuum between the frames and the table, and also between the two frames. The table is drilled with exit holes for the evacuation of air.

Moulds

In foundry work, the patterns are of prime importance, for they are responsible for the shape, accuracy and finish of the resulting casting. Their production is, therefore, time-consuming, and complicated patterns are extremely difficult to design and manufacture. But in vacuum forming, the mould form or pattern need not be elaborate, expensive or difficult to produce. To take a simple example: a rectangular piece of wood with its four edges slightly tapered—'draught' in pattern-making terms—may be used to form a container of precisely the same shape. In general practice, moulds are made from wood, plaster, metal-filled resins, thermosetting resins and metal castings, and this gives scope for the choice

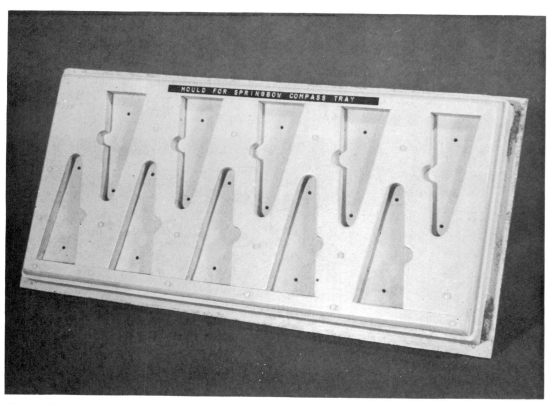

MOULD FOR SPRINGBOW COMPASS TRAY

Plate 46   The mould for a springbow compass tray

of material that is available or that can be shaped most easily
and effectively for the task in hand. Some school moulds, all
made in wood, some with plaster filling, others simply sawn
and planed to shape, are shown in plates 46, 47 and 48.

The surface finish of the mould is very important, for once the
sheet material reaches its shaping temperature and is brought
down onto the mould and formed it will have in its surface
every detail from the mould, including all defects. All mould
surfaces should be smooth, the grain filled if made of wood,
and the surface sealed with heat-resistant paint or varnish.

Draught or taper is important to enable the component to be
removed from the mould once the forming and cooling are
complete. It should be mentioned, however, that on thin sheet
material the moulding is flexible enough even after forming

121

to be prised or sprung off the mould without damage resulting.
As a general rule, all vertical faces of a mould should have
draught as follows:

(a) Male moulds—approximately 5°
(b) Female moulds—approximately 1°

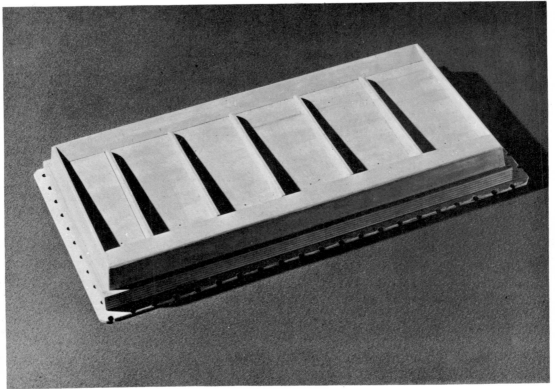

Plate 47   *The mould for a decimal coinage tray*

Mould Venting   This is a most important aspect of successful mould design.
To ensure that all air is evacuated from between the mould
and the sheet material as quickly as possible—i.e. before the
sheet temperature falls below its forming temperature—holes
must be drilled in the mould surface, especially where air is
likely to become trapped. Small depressions or cavities in the
mould surface could trap pockets of air at the forming stage
and thus prevent a positive, accurate impression coming off

the mould. The size of hole should be quite small—approximately 1 mm diameter—for a large hole would cause the material to be pulled down into it during forming and might even puncture the plastics sheet. These small holes are usually counter-drilled from underneath the mould with a larger drill—say 4 mm or 6 mm diameter—in order to avoid any restriction on the outflow of air. Part of a contour map with the venting holes clearly visible is shown in plate 48.

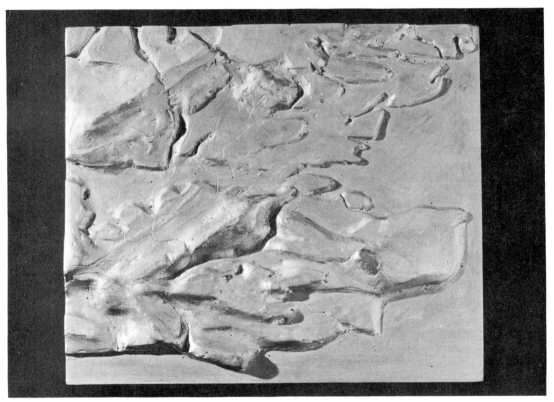

Plate 48   *The mould for a contour map (note the positioning of vent-holes)*

Forming Techniques

There are five basic techniques in vacuum forming and these are:
1 Female forming
2 Box forming

123

3 Drape forming
4 Bubble forming
5 Plug-assisted forming

The first three—female, box and drape forming—are straightforward processes and can be attempted at school level. The last two, however, are more advanced and really apply to industrial production.

Female Forming (fig 33)

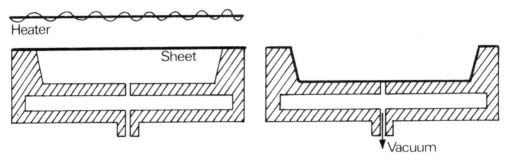

Fig 33

This is the forming of a sheet of material into a hollow of female-mould form. It is a simple process but should be used only on shallow formings. Remember that female moulds need very little draught on the vertical surfaces.

Box Forming: This is illustrated in fig 32 at the commencement of the chapter, and it will be noted that the mould is positioned in a box-like container—hence the name. The softened sheet is drawn down into the box and over the mould. This technique consumes more material per article than female forming, but it gives a slightly better uniformity of wall thickness.

Drape Forming (fig 34)

This process involves draping the softened sheet material over the mould form—partly shaping it—before evacuating the air. To achieve this, the sheet is heated above the mould and at the shaping temperature, the sheet is brought down or the table supporting the mould is brought up, thus causing the draping of the material to take place. Forming then proceeds

124

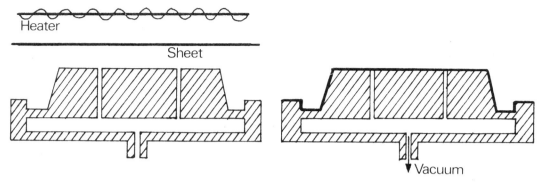

Fig 34

by evacuating the remaining air. Drape forming may be more economical on the material than box forming and the wall thickness uniformity is about the same.

Possible Faults in Vacuum Forming

1 **Blowholes:** These may appear if material becomes overheated and loses its strength when it is stretched over the mould or subjected to vacuum.

2 **Splits:** Here the material splits—especially on sharp edges or corners. Radius off such sharp edges and corners on the mould.

3 **Uneven Forming:** The finished forming may lack uniformity of definition, due to uneven heating or inadequate venting in the areas concerned. A check should be made of both the heater layout, to see it gives sufficient coverage to the whole area of the sheet, and the venting of the mould.

School Applications

There will be many opportunities for co-operation between various departments such as Art, Mathematics, Geography, Metalwork and Technical Drawing, who all have requirements related to vacuum forming. From this liaison will stem an appreciation of what this technique can do to solve a variety of design problems. Contour-map work, drawing-instrument trays and a tray to store practice decimal coinage (plates 48, 49, 50) were all produced in school by this technique.

125

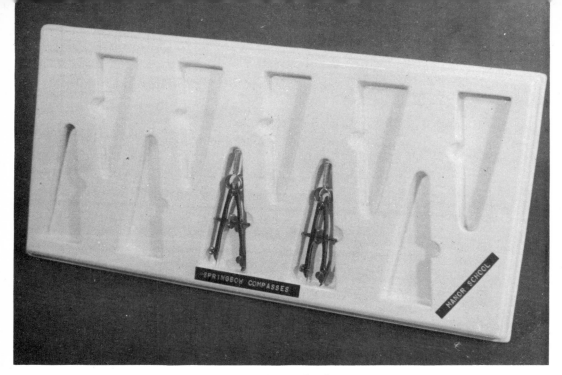

Plate 49   *The finished tray for springbow compass*

Plate 50   *The finished tray for decimal coinage*

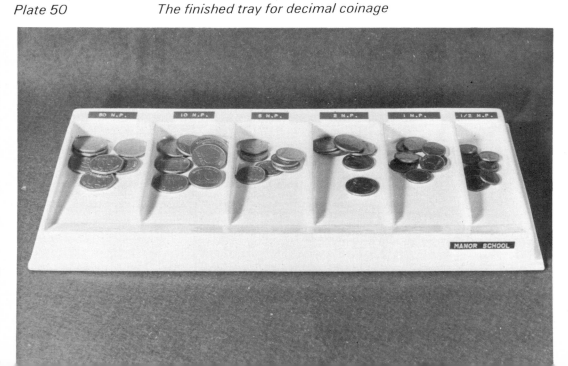

Vacuum Forming in School

Materials: For a small hand machine sheet materials up to approximately 1·5 mm such as PVC, polyethylene, cellulose acetate and polystyrene are satisfactory.

Equipment: Mould form, vacuum forming apparatus.

Procedure
1 Cut the sheet material to size of the clamping frame and clamp down firmly.
2 Switch on the heater and offer the sheet up to the heater. Some machines may have the heater horizontally above the sheet.
3 Position the mould on the table of the machine.
4 When there is evidence that the sheet has become sufficiently plastic, bring the clamping frame quickly onto the mould and evacuate the air.

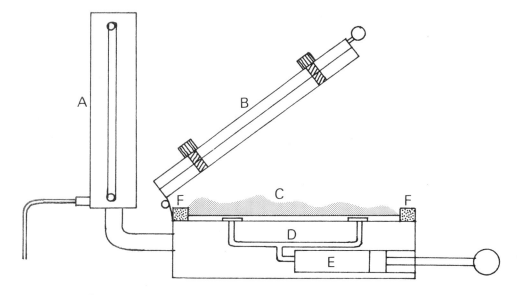

A Heater unit
B Frame for clamping the sheet
C Mould form
D Tubes connected to the pump to draw air from the mould area
E Hand pump
F Sponge rubber for sealing the edges

Fig 35

5 Observe how the forming progresses and, if necessary, give some additional heat to any part of the sheet that may require it.

6 Bring the vacuum forming off the mould and remove it from the clamping frame. Trim off the waste material.

A cross-sectional view through a small hand-operated vacuum forming machine which embodies the drape forming principle is shown in fig 35.

Economy Note: In shallow mould work the sheet may be used more than once by reheating, which causes it to return to a flat surface. This tip is also useful if a moulding is not quite satisfactory.

2 Injection Moulding

Introduction

In 1966 about one-sixth of all plastics materials used throughout the world were processed by injection moulding and this process must feature even larger today. Many of the plastics articles we handle from day to day have been shaped by injection moulding.

Injection moulding is basically a very simple process. A plastics material is heated in a cylinder and becomes 'plasticized'; it is then forced under considerable pressure into a mould where it solidifies; the component is finally removed and the mould made ready for the next moulding. It is very similar to compression moulding in several ways—a mould is used, heat is applied and great pressures are developed in the forming of the material. Injection moulding, however, is concerned with thermoplastics materials and compression moulding is normally associated with the thermosetting plastics.

The history of the technique goes back almost a hundred years to when the practice of pressure die-casting of metals was tried out on a plastics material. Pyroxylin, the first plastics material, was used, and experiments involving heat and pressure produced the first ivory-substitute billiard balls. Much of the process developed then is still used in today's modern machines.

The capacity of a machine is given by quoting the weight of the largest moulding it can produce. Moulding machines vary in size from small hand machines delivering components weighing only a few grams each shot to large production types having a capacity of 25 kg and over each shot.

Plate 51   *A fully automatic injection moulding machine*
Courtesy of ICI Plastics Division

Main Parts of an Injection Moulding Machine

1 Feed Hopper: A container in which is stored the plastics moulding material. It is directly linked to the cylinder of the machine and it can supply a measured quantity of material as required.

129

2 Cylinder: A hollow tube in which runs either a plunger or a screw, both of which can carry along the material to the heating or plasticizing zone.

3 Heater or Plasticizing Zone: An electrical heating element or elements are wrapped around the cylinder supplying sufficient heat to plasticize the moulding material. Usually at this point in the cylinder, there is a torpedo or spreader—this device pushes the plastics material towards the walls of the cylinder, thus ensuring that it comes into complete contact with the heat supply.

4 Nozzle: This is the orifice through which the material is squirted into the mould cavity. It fits the hole in the surface of the mould called the sprue; it usually has built into it a shut-off valve to prevent loss of material.

5 Pressure: This can be supplied to the screw or ram, mechanically or by hand. Most machines are mechanically

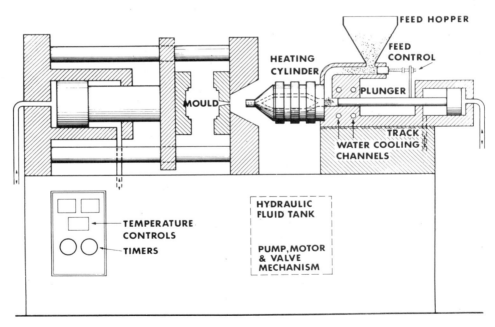

Plate 52

Sectional drawing of a plunger injection moulding machine
Courtesy of ICI Plastics Division

130

operated but small bench-type machines can be hand operated. Mechanical operation is usually by hydraulic pressure, hand operation by a capstan handle fixed to a rack and pinion feed mechanism.

The Moulding Cycle

1 Assembly, closing and clamping of the mould in the machine
2 Injection of the material
3 Dwell time—plunger/ram held down
4 Ram withdrawn
5 Setting time for the material
6 Unclamping and opening of the mould parts
7 Extraction or ejection of the component or components
8 Repeat . . .

The above cycle applies on a school level as well as in the industrial situation. Industry, of course, is concerned with time spent on all the above stages in the moulding cycle, but in a school this is of little or no consequence. However, several points must be investigated, explained and remembered:

(a) In the assembly of the mould, check that no remains of the previous moulding are present on any of the mould surfaces. Small particles can prevent the mould closing properly with consequent leaking at the joint line. Clamp the mould very tightly to ensure there are no gaps between surfaces.

(b) Inject the material swiftly and steadily—no pauses, as this may cause premature hardening in the mould sprue, runners or gate.

(c) The dwell time is very important—as the pressure is maintained, the material injected is unable to creep back towards the sprue and hence a better-quality component, free of shrinkage, hollows or voids, is produced.

(d) Ram withdrawal—no problems.

(e) Time must be allowed for the material to freeze, i.e. go hard. Premature opening of the mould may result in the deformation or fracture of the component.

(f) Opening the mould requires care. Over-enthusiastic use of hammers, screwdrivers and knife blades will result in the mould being damaged. Some moulds can be made with slots in the joint lines to accommodate a screwdriver blade. If the mould requires some sharp taps to open it, only a mallet made of nylon, wood or fibre should be used.

131

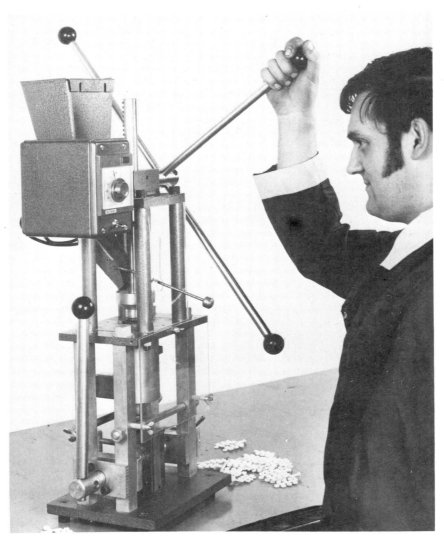

Plate 53   *A hand machine for injection moulding*
Courtesy of Small Power Machine Co. Ltd

(g) Ejection of the component—this can be hot and possibly still sufficiently plastic to become deformed by heavy handling.

The Mould for Injection Moulding

The success or failure of injection moulding hangs on the mould both in its design and its manufacture. Compared with those moulds in commercial use today, moulds made in a school

132

Plate 54   *Two-part moulds for a draught and a door stop machined in school*

Plate 55   *A three-part mould for a door knob machined in school*

will be simple and within the scope of boys working in the school engineering workshop. No real advantage can be gained by the importation and use of highly sophisticated moulds. Such moulds would, of course, demonstrate the process, but much more value can be gained from injection moulding if the mould is designed by a boy or a group of boys, in consultation with the teacher, and then produced by them in the workshop. A sample of such moulds is shown and all have been proved to be well within the ability of pupils in secondary schools.

A mould usually consists of two or more parts, these parts having cut or machined in them the shape of the component to be moulded. All surfaces should be very smooth and polished, especially the faces of the parts of the mould that fit together. Any gaps on the split line will result in the mould leaking and producing unwanted 'flash'. On one of the split lines will be positioned the hole through which the plastics material may be injected. This is known as the sprue and fits the nozzle of the machine. The sprue is joined to the mould cavity by a gate, and if the mould is a multi-impression type, the sprue feeds a runner off which are gates to each individual mould cavity. The gate is usually a narrow constriction, allowing molten material through but easily broken off when the component is finally removed. To extract components, an ejector pin or pins can be incorporated in the mould so that, when it is unclamped and opened, the pins can be tapped and the component ejected.

As with pattern making for casting and foundry work, certain basic points of mould design should be adhered to. Sharp angles and corners should be avoided as these create inherent weaknesses. Mould cavities must have adequate draught or taper on all vertical surfaces so that the moulded component can be extracted easily. Sprues, runners and gates should be round in section so that the material has uninterrupted flow and they should be large enough to allow the cavity to be filled quickly and completely.

Types of Gates   1 **Pin Gate:** This type is for small, thin-section mouldings. It allows easy break-off of the sprue and a minimum of finishing at that point.

134

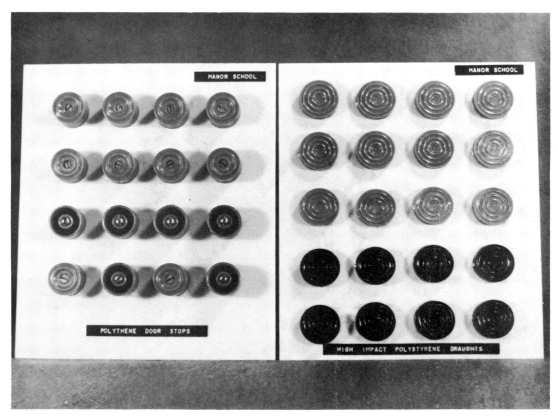

Plate 56   *A selection of mouldings*

2 Side Gate: Used when the plastics material can be fed into the edge or side of a component.

3 Sprue Gate: No restriction with this one—just the sprue connects to the component. Used on large mouldings— buckets, baths and bowls.

4 Fan Gate: This is a gate which spreads out on the cavity side of the mould so as to gain a good broad distribution of moulding material over a large area.

5 Tab Gate: Used when large areas, especially in transparent materials, are moulded. In this situation, flow marks are likely to occur, thus spoiling the moulding. The tab allows turbulence to take place, obviating what is called 'jetting' on the surface of the component.

135

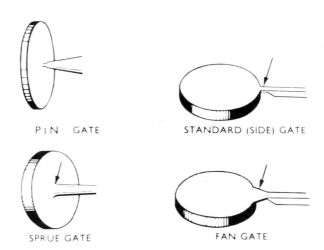

PIN GATE

STANDARD (SIDE) GATE

SPRUE GATE

FAN GATE

Plate 57

Four common gates

Moulding Faults and Remedies

1 **Flashing:** This is where fluid material has spilled out from the parting line(s) of the mould.

Remedy: Check parting line surfaces for cleanliness.
Check mould vice pressure.
Check mould alignment.

2 **Short Mouldings:** Mouldings that are incomplete due to insufficient material or material reluctant to flow.

Remedy: Check cylinder temperature.
Increase injection pressure.
Check gate, sprue and runner sizes and enlarge if necessary to increase the flow rate.
Inject material more quickly.

3 **Voids:** Cavities or spaces that form as the plastics material cools, due to there being insufficient material in the mould.

Remedy: Keep the plunger pressure on longer.
Reduce the temperature of the material.
Enlarge the gate to prevent premature solidification.
Check that there is sufficient material available to fill the mould and to maintain pressure during dwell.

4 **Weld Lines:** This comprises evidence of where the material has flowed different ways after entering through the gate and has subsequently joined up in the moulding.

136

Remedy: Check the mould temperature.

Check the temperature of the material.

Speed up injection.

Increase the pressure on the plunger.

Re-position the gate so as to cut down the distance the material has to flow.

There are many other faults that can occur, but those shown above are the ones that commonly appear in school work.

Suggested Machine Cylinder Temperatures

Material	Temperature, °C
Polyethylene	190–220
Polystyrene	200
Polypropylene	250
Nylon	190–220

Suggested Mould Temperatures

Material	Temperature, °C
Polyethylene	30–50
Polystyrene	30–50
Polypropylene	50–60
Nylon	85–100

Making Injection Mouldings in School

Materials: Polyethylene, polystyrene, polypropylene, nylon, PVC, ABS

Equipment: Injection moulding machine, moulds, face guard, asbestos gloves, mallet that will not damage moulds, table knife.

Procedure

1 Switch on the heater unit.
2 Load moulding granules into the hopper of the machine. These feed by gravity into the cylinder where a ram or screw carries the material to the heater zone.
3 Prepare the mould and clamp into the vice of the machine.
4 Check that the heater temperature is correct for the material.
5 Inject the material into the mould and maintain the pressure for a short time. Allow freeze time for the material in use.
6 Withdraw nozzle of machine from the mould and top up the cylinder with more material.
7 Unclamp the mould, open the parts carefully and remove the product.

Safety Notes

1 Care should be taken to see that any moisture present in the granules has been driven off by a pre-drying stage—particularly nylon.
2 The machine should have an adequate guard around the nozzle area to prevent injury by any possible blow-back of hot material.
3 The operator should wear a face guard for protection in the event of a blow-back from the hopper end of the cylinder. On many horizontal machines this is not liable to occur.

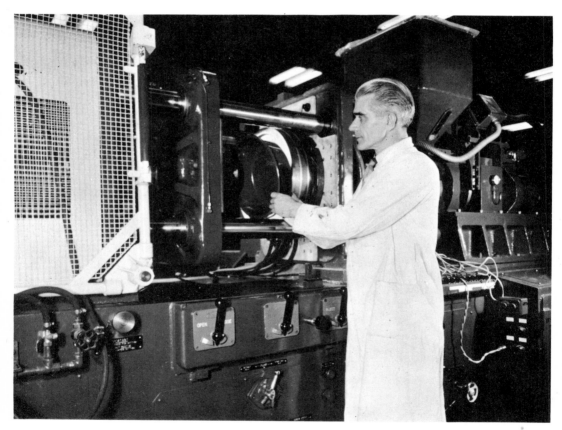

Plate 58　　　*A domestic bowl being injection moulded in 'Alkathene', ICI polyethylene*
Courtesy of ICI Plastics Division

138

3 Welding of Plastics

Introduction

Welding consists of the joining of two or more pieces of the same plastics material using, primarily, heat to produce a joint or union of a very similar composition. We can immediately compare the welding of metals—which may have been carried out in the school workshop—with the welding of plastics. Usually metals of the same type are welded together; heat is used from one of several alternative sources—oxy-acetylene, electric arc or high frequency. The joint or union is made up essentially from the same material as the pieces of metal being joined. Therefore much of what is relevant to the welding of metals will be equally relevant to the welding of plastics, and the equipment and techniques used are very similar.

Methods of Welding

The following are the most commonly used techniques today for the welding of plastics materials:

| Heated-tool welding | Friction welding |
| Hot-gas welding | High-frequency welding |

1 **Heated-tool Welding** is used essentially in the sealing of films and sheets. The common polyethylene bag used in such a variety of jobs and applications today is sealed at one end by this method.

2 **Hot-gas Welding** is exactly akin to the technique of welding of metals—a torch and a filler rod are used in order to join or seam up surfaces or to fabricate structures.

3 **Friction Welding** involves using one of man's oldest techniques, that of generating heat by rubbing two objects together. Frictional heat can be a nuisance in many practical tasks, but it can be put to good use in the joining of a range of plastics.

4 **High-frequency Welding** is the inducement of heat within the plastics material by the application of dielectric heating. Uniform heating results and hence a better-quality joint is produced. This technique is used for sealing containers, clothing seams and soles of shoes; it is also used in the sealing

of articles where direct heat would affect the contents sachets of shampoo or packeted foodstuffs.

In schools, friction welding and heated-tool welding can be attempted with the minimum of equipment and both present very few difficulties to pupils. Hot-air welding requires specialist equipment but it can certainly be attempted at school level with much success. The building of hot-air welding equipment could form the basis of a project. High-frequency welding does require very specialized equipment, which is expensive, and a good electrical background would be needed to build the apparatus. For practical purposes, therefore, only heated-tool welding, hot-air welding and friction welding will be covered in detail.

Suitable Materials
From what was said in the introduction it may be deduced that the only suitable materials for welding are the thermoplastic varieties. A plastics must be of the heat-softening type to respond to a heated tool, hot-gas stream or friction. Most thermoplastics can be welded provided they are not inflammable or they do not decompose when their softening temperature is reached: two examples of plastics that cannot be welded are cellulose nitrate, because of its high degree of inflammability, and PTFE which decomposes on passing its softening temperature.

Chart of the Common Weldable Plastics

Material	Heated-tool Welding	Hot-gas Welding	Friction Welding	High-frequency Welding
Polyethylene	●	●		
Polypropylene	●	●		
P.V.C.	●	●	●	●
Nylon	●	●	●	●
Cellulose Acetate	●		●	●
Cellulose Acetate Butyrate	●		●	●
Acrylics	●	●		●
Polystyrene	●	●	●	

140

Heated-tool Welding

This probably is the most widely used welding process and is applicable to the sealing of films and sheets. The basic principle involved is the heating of the surfaces to be joined by a heated roller, edge or plate and the maintainance of this heating until the material softens. The softened parts must then be pressed together and the pressure maintained until the plastics material cools and the joint solidifies.

In its simplest form heated-tool welding can be carried out with a soldering iron or similar-shaped tool, run along the edge or joint to be sealed and followed up by a roller to press the edges together. However, joints produced by this relatively simple, basic method will neither have a very good finish nor necessarily possess a uniform strength throughout the length of the seam or joint.

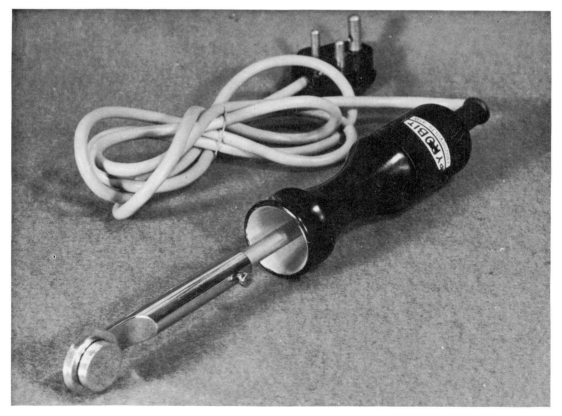

Plate 59   *An Acru tool for welding*

A better type of tool is the combined heated roller type that can be made in the school workshop, or purchased. An example of a commercially produced tool is shown in plate 59. Below the handle a heater element is contained in a tube and this heats the roller holder which fits the tube very closely; the roller revolves freely on a spindle fitted with a knurled nut and this roller can be changed for other patterns for different work. The temperature of the roller can be controlled to a limited degree by sliding the roller holder up or down the tube, thus exposing it to more or less contact with the heater element. A 45-W heater operating on 230/250 volts would be quite suitable for sealing polyethylene sheet up to 1000 gauge. The tool is also shown in use, sealing the end of polyethylene lay-flat tubing and using a clamping jig to hold the material steadily as well as guide the roller—plate 60.

Plate 60   *The Acru tool in use*

Use of Barrier Film: It can be readily understood that when a heated tool is used the material being welded can and in fact does melt as the welding proceeds and deposits become stuck to the roller. To avoid this happening a barrier film should be used between roller and material. This film is a polyester film —'Melinex' produced by Imperial Chemical Industries Ltd— and it enables welding to be carried out with the minimum of trouble. Polyester film has a high melt point and can be used many times over, although with prolonged usage it does become stretched and distorted. It can then be discarded. A suitable gauge of polyester film for welding polyethylene sheet has been found to be 200 gauge.

Precautions

1 Check that the material is clean and free from grease, oil and even finger marks.
2 Try out a test piece, initially, to ascertain the correct welding temperature for the material being used.
3 Do not overheat the material, for this will result in a weak joint or seam, also decomposition or burning of the material.
4 Use a jig wherever the work allows it. This will ensure straight weld lines and stability of the material being welded.

Safety Notes

1 Before making up or using any electrically heated sealing tool, check that it is completely safe and properly earthed.
2 Design a holder for the tool so it can be supported safely when it is not in use.
3 Switch off the heater when the tool is not needed—this prolongs element life and is safer for all concerned.

Hot-gas Welding

This process can also be called hot-air welding, for a heated air or gas jet, nitrogen, is commonly used. Heat is applied to both the surfaces to be welded and the filler rod. As in the welding of metals, a filler rod is used to fill in the space left between the surfaces. The rod is softened and so lies in the junction between the surfaces, producing a joint which, if welded correctly, has the same strength as the materials joined.

The equipment consists of a torch which has an electrical heater element situated in it. In some models, the gas or air supply is fed in separately from a compressor or cylinder of gas; on others the air supply is contained within the torch in the shape of a small blower motor. Torches are also manufactured to operate with a combustible gas supplying the heat source together with the compressed gas or air supply. Using any of these methods, a hot air or gas stream is created. Plate 61 shows a typical small torch suitable for use in schools; this can be connected to a paint-spray compressor or the brazing-hearth compressor in the workshop. A flowmeter, which is a device giving the rate of flow of air or gas in litres per minute, can also be seen. This is a useful accessory for continuous accurate work on a production-line basis but it is not essential for school work. The torch can have a variety of nozzles fitted to it to do a range of work, but again, at school level, a choice of three or four nozzles would be ample.

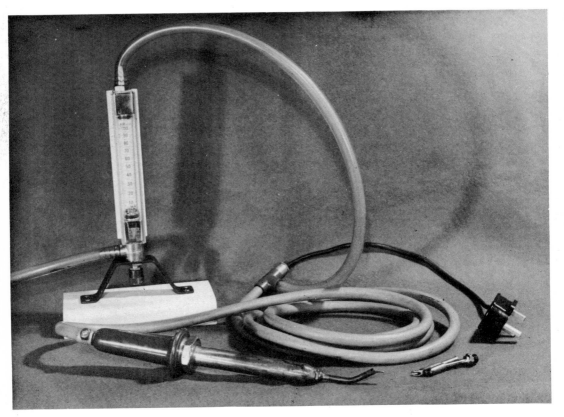

Plate 61   *A hot-gas welding tool*

Welding conditions for plastics materials vary according to their physical properties. Some materials melt completely whereas others soften and no more. Successful welding requires that both material and filler rod reach the same temperature together and then, with a degree of pressure exerted on the filler rod, are made to fuse together. The hot stream of gas or air must be at the correct temperature for the material being used and this temperature can be controlled by varying the supply of air or gas to the torch. The usual range of temperature produced by a hand torch is from approximately 200°C to 500°C.

Procedure

1 The material to be welded must be clean and prepared for the weld. Reference to fig 36 will show how the edges to be welded are bevelled at an included angle of between 60° and 70°. This 'vee' is subsequently filled with welding rod of exactly the same type of material. The welding rod also must be clean—rub clean emery cloth over the length of the rod to be used.
2 Pre-heat the end of the material where the weld is to commence.
3 Attach the end of the filler rod to this point and with a back-and-forth movement of the torch heat both the rod and material together.
4 Downward pressure in the direction of arrow X will tend to create a forward movement in the direction of arrow Y. Make certain that the feed of the filler rod is vertically downwards.
5 Any parts of a weld that become overheated and burnt should be scraped clean and re-welded.

With practice on scrap pieces of material a degree of expertise can be achieved, and skill and technique can develop in quite a short time. One important point is that dissimilar materials cannot be welded; PVC and polyethylene, for example, do not weld together. It is also important that the grade of filler rod and sheet are identical so that the melting points of the two are identical.

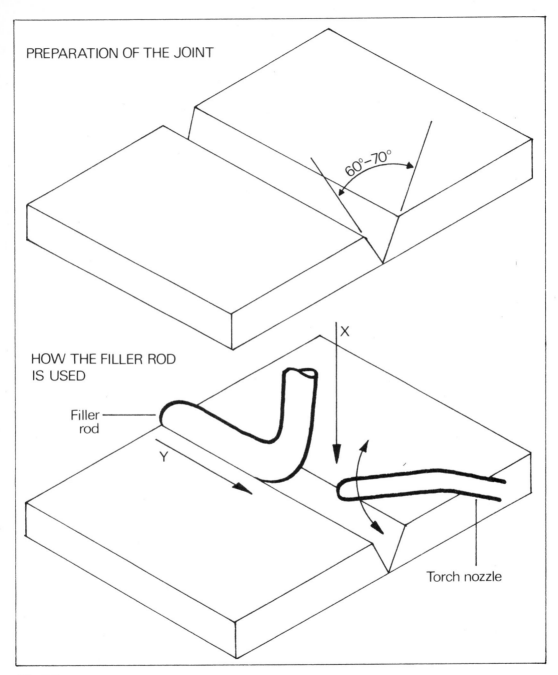

PREPARATION OF THE JOINT

60°–70°

HOW THE FILLER ROD IS USED

Filler rod

X

Y

Torch nozzle

Fig 36

146

Recommended Weld Joints: The following are common welds and are considered suitable for thermoplastic materials:

1 Butt weld (fig 37)

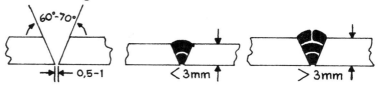

Fig 37

2 Double-butt weld (fig 38)

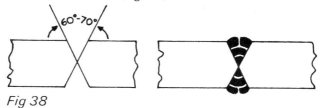

Fig 38

3 Fillet weld (fig 39)

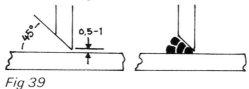

Fig 39

4 Double-fillet weld (fig 40)

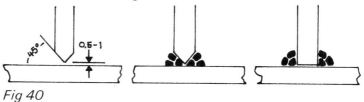

Fig 40

5 Corner weld (fig 41)

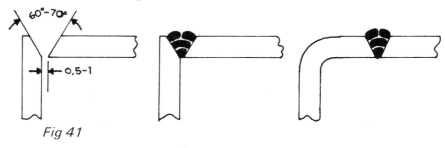

Fig 41

147

Industrial Applications

1 The fabrication of containers and structures.
2 The making of pipe connections and in plumbing generally.
3 Joining PVC coated cloth using a tape instead of a filler rod.
4 Film welding using just the hot-gas stream.
5 Butt welding tube and extruded forms to produce long lengths.
6 The forming and bending of plastics where localized heat is required.
7 Roofing installations and awnings.
8 Reservoir linings.
9 Jointless PVC flooring.
10 The lining of tanks and containers which may contain chemicals or corrosive liquids.

It is worth mentioning here that a supply of hot gas or air can solve many problems apart from its use for welding purposes: it can be used to dry out a mould cavity prior to casting; it can be used to dry off a painted surface; it can be used to soften acrylic sheet for bending and shaping in a particular spot and so on.

School Applications

Besides giving opportunities for very useful welding practice, possible applications could be:
1 The fabrication of containers usually made in other materials.
2 Art work—three-dimensional structures, collage.
3 Repairs.
4 Drawing-office models.
5 Model-boat hulls.

Safety Notes

1 The air or gas stream temperature can be anything between 200°C and 500°C, and this jet could be extremely dangerous where pupils are concerned. An adequate torch support is needed so it can be held safely pointing away from the operator whilst it is not in use. Some manufacturers of torches build a support into them.
2 Protective gloves should be worn.
3 *Do not* switch off the air or gas supply *before* switching off the element. Element burn-out will be caused by this, for the electrical element is designed to operate with air circulating around it. In fact air or gas should pass over the element to

148

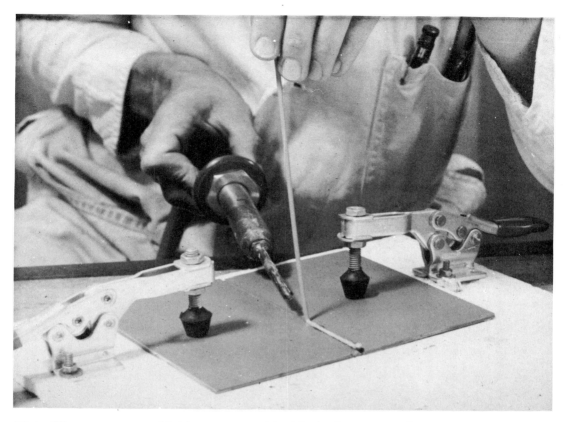

Plate 62   *Making a vee weld with the hot-gas tool*

cool it after it has been switched off. Again, some manufacturers of torches have built in a safety device to avoid this sort of damage to the element.

Friction Welding   Two surfaces rubbing together produce friction or heat, and unless this friction is checked, more and more heat will build up at an increasingly faster rate, depending on the material's heat conductivity ability. If the two surfaces rubbing together are thermoplastics and the heat is allowed to build up, a point will be reached where softening of the surfaces occurs and at this point the two will adhere or weld. Provided one of them is then allowed to move with the other, the two will remain

149

joined. Incidentally, friction welding can occur in metals where no lubrication is present and this presents the engineer with a problem he would prefer not to have!

Some Advantages of Friction Welding
1 Welds can be made quickly.
2 The strength can be equal to the parent material.
3 Thermoplastic material can be joined to a thermosetting material if this is required by a 'keying in' process, not a fusing of surfaces.
4 Gap filling takes place if the materials being joined are not quite flat. Good surface contact results therefrom.

Tests have been carried out on cemented and friction-welded joints using acrylic material and the welded joints show themselves to be approximately twice as strong as a cemented joint. Light transmission, however, is the same in both examples.

Materials that can be friction welded
1 Cellulose acetate
2 Cellulose acetate butyrate
3 Nylon
4 Acrylics—to a limited extent
5 PVC
6 Polystyrene

Equipment needed: The usual workshop equipment is quite sufficient for the friction welding of plastics—pillar drills or centre lathes—or any item of equipment where one piece of the material can be turned or rotated and the other piece held temporarily stationary up to the moment of a weld taking place, when it must be able to turn or rotate with the first piece.

Speeds for Welding: Generally, speeds should be high, but other than that it is very difficult to list speeds relevant to all materials and their diameters. The best approach is to start the material on a medium speed, bearing in mind that it can be increased until welding is obtained. Pressure also plays an important part and here, too, trying out a material will show what is a suitable pressure to apply. Thermoplastics generally are poor heat conductors so it is difficult to overheat any material. Pressures of between 100–300 lb/in^2 are commonly used in friction welding.

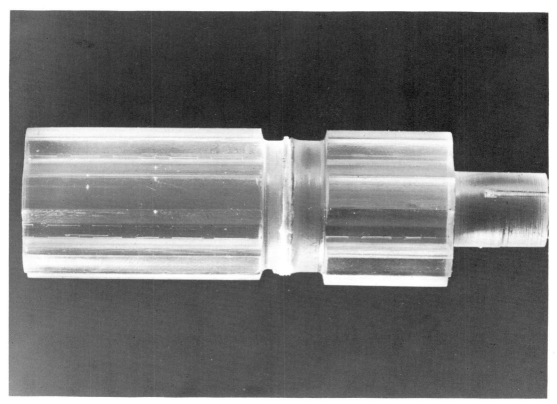

Plate 63   *A friction-welded cellulose-acetate rod*

Procedure

1 The surfaces do not have to be actively cleaned because any dirt or surface film will be squeezed out very quickly.
2 Clamp one piece of material in a chuck in the centre lathe or drilling machine.
3 Set up the other piece of material so that it can be released on welding of the surfaces—a possibility here for a project on a variable friction-controlled chuck or fixture that could be produced in the school workshop.
4 Select a speed suitable for the material—prior experimentation with scrap can ascertain this.
5 Commence welding—the correct temperature will be seen to be reached when the material visibly flows.

151

6 Remove the welded article and break off the flash from the joint area. This should come away easily as it is usually hard, brittle and thin in section.

Note: Only similar materials should be friction welded as described because the different softening points present in two different materials could make the welding difficult or even impossible.

Safety Notes
1 Ensure that a system has been devised in such a way that the piece that is stationary for the first part of the welding cycle can begin to rotate at the moment of welding **without endangering the operator**.
2 Start welding with a lower speed than is thought necessary—get the feel of the material.
3 Wear goggles on occasions when brittle material is being welded—it could shatter into dangerously small pieces if the pressure is too great.

4 Use of PTFE

Introduction

Polytetrafluoroethylene—PTFE in its accepted abbreviated style—is considered in this context in a dispersion form, that is in a liquid or fluid form, so that it may be applied as a coating to a metallic or non-metallic surface. Usually a metal is the chosen base on which to coat, but wood, ceramics and some plastics can be treated as well.

Why PTFE is used as a Coating

To answer this it is necessary to know a little about the material itself and its physical properties. Firstly, it belongs chemically to the same family as PVC, but when the production of the material is considered, the process is an elaborate one and hence the cost of the material is considerably higher than PVC. Because of its relatively high cost, it has become a specialist material, employed when the electrical, chemical or mechanical engineer requires high standards of operation, or where there is a problem which could not be solved by the use of another material.

Chemically, PTFE is completely inert—that is, it is not attacked. It can resist high and low temperatures with little or no loss of its properties—it can operate satisfactorily up to approximately 300°C and down to −200°C and still retain its flexibility. Electrically, it has extremely good insulating properties; for example, a wire of 0·024 in diameter with a coating of PTFE of 0·0014 in thick has a dielectric strength of 2080 volts! It has an extremely low-moisture-absorption factor and it is therefore very stable when dimensions are important. It has the lowest coefficient of friction of any known solid material and lastly its surface is non-sticking to almost all materials, even glues and adhesives.

Because of these outstanding properties, its uses are many and various: PTFE is used as a liner in pipes and containers which are to hold highly reactive chemical substances; pipes and hoses to carry such materials as liquid air are lined with it; bearing surfaces are coated with it or fabricated from it in the solid form where oil or lubricants cannot be applied; surfaces that are to be in contact with sticky materials or foodstuffs are coated with PTFE; cooking utensils—pots, pans, dishes, baking trays and tins—are coated to facilitate removal of the cooked food. Where electrical insulation is required at extremes of temperature PTFE is used and not the more common material polyethylene.

Manufacturers of PTFE Dispersions

Well-known trade names are synonymous with PTFE coatings; 'Fluon' is the trade mark of Imperial Chemical Industries Ltd; 'Teflon' is the trade mark of El du Pont de Nemours & Company Inc. PTFE dispersions are a suspension of the plastics in water, the water being the vehicle by which it is applied. Common colours of the finish coat are white, grey and black.

Grades of Dispersion

Different grades or types are used in coating processes, depending on the underlying material, but at a school level, metallic surfaces will invariably be the base material and this can be coated with either:

(a) A one-coat finish
(b) A primer and finish coat

153

The following are grades of Imperial Chemical Industries Ltd 'Fluon' PTFE dispersion:

(*a*) 'Fluon' GP1—This is suitable as a general-purpose coating agent.

(*b*) 'Fluon' SC1—This has to be mixed with 'Fluon' SC2 acid to form a primer which is sprayed onto the surface prior to a finish coat of 'Fluon'. The SC1 contains a percentage of PTFE.

(*c*) 'Fluon' SC2—This is the acid for SC1, and when mixed with SC1 forms the etching part of the primer coat. The metallic surface requires this etching so that the coating adheres properly.

(*d*) 'Fluon' SC3—A coloured version of SC2, and when mixed with SC1 produces a one-coat finish.

(*e*) 'Fluon' Clear and Black—Top-coat finishes of 'Fluon' which are used on top of the 'Fluon' primer.

Applications

The coating of PTFE on cooking and kitchen utensils is the best-known domestic use of PTFE and in the school workshop this is the obvious application. Coating can be carried out on cast hollow-ware—aluminium alloy or cast iron—although it is more often applied to discs of metal such as frying pans and saucepans which are subsequently spun to shape. The facilities which exist at school-workshop level will decide the approach to the coating process. The following description of coating is based upon the coating of a cast aluminium-alloy utensil, cast using the foundry facilities within the school. The grade of dispersion selected is the primer and finish coat—primer (SC1 plus SC2) and black top-coat finish.

Before proceeding with any work using PTFE in dispersion form, consult Imperial Chemical Industries Ltd Technical Literature F.10—'Medical Aspects of Polytetrafluoroethylene'.

Equipment

1 De-greasing facilities.
2 An oven capable of supplying a controlled temperature of 90°C, Oven No. 1.
3 An oven capable of supplying a controlled temperature of 380–400°C, Oven No. 2.
4 Spraying facilities.
5 Extraction facilities for both spraying and exhaust fumes from the ovens.

Plate 64   *A frying pan made and coated with PTFE in school*

Procedure

De-greasing: The surface to be coated must be de-greased thoroughly, by immersion in a suitable solvent such as 1 : 1 : 1-trichloroethane, and allowed to dry off fully before proceeding to the next stage. On a cast piece of hollow-ware the surface must be free of porosity, for these small pockets or pinholes on the surface will harbour grease, moisture and solvent deposits, and the subsequent coating will be affected by it.

Mixing SC1 and SC2: Measure 60 parts by weight of SC1 'Fluon' primer against 40 parts by weight of SC2 acid; for example, 60 g of primer and 40 g of acid. The SC2 acid should be added to the SC1 primer slowly, stirring as the two mix. On no account add primer to acid as this will cause the primer to coagulate. When the two have been mixed

155

properly together, the mixture will have a shelf life of about two weeks, at a temperature of 20°C (68°F).

Spraying: Two methods of applying PTFE dispersion exist—spraying and spin coating. Where a formed/shaped article is to be coated, spraying is the only method of application available; a flat disc can be spin coated and then subsequently formed into its final shape. This cast workpiece requires the spraying technique. A gravity-fed spray gun as commonly used in paint spraying is suitable. Nozzle diameter needs to be between 0·060 in and 0·080 in and an air pressure of about 20 lb/in² is required. A suitable gun is produced by Aerograph de Vilbiss, Model MPS. The manufacturer's address can be found in Appendix 3.

As with all spray work, mouth and nose protection should be worn and extraction fans should be in use to dispose of all atomized material. Spraying of both primer and finish coat has to be carried out to give a depth of coating of less than 0·001 in. This sounds quite a difficult task, but a small amount of practice and observation of colour change will assist a great deal. Too thick a coating will cause 'mud cracking' to develop as the coat dries and this is the reason why PTFE dispersion cannot be brushed on—the depth of coating is really critical.

There are two final points that will assist in judging whether the primer and finish coats are thick enough:
(a) The primer should have a light to medium yellow colour when it has been sprayed on to the correct thickness.
(b) The finish coat should cover the primer, now a light brown after the drying stage, but should not run down vertical or semi-vertical surfaces.

Drying: With the primer applied, drying has to be carried out and this should be done in the oven set to operate at 90°C. What happens during this drying-out phase is that the water base of the PTFE dispersion is dried off, leaving the PTFE only on the surface of the metal. If the drying off is carried out too rapidly, the water will boil and cause 'volcanoes' to appear and hence disfigure the finish.

Cooling: After drying for approximately 15 minutes—but this will depend on the size of the object—a cooling period takes place to get the metal back to the room temperature. On no account should the primed surface be touched. The primer will have turned a dark yellow/brown colour at this point.

Spraying the Finish Coat: As PTFE dispersion is a water-based material, the spray-gun parts can be washed out after the priming has finished. This can be done under running water, with the air supply running, and the equipment can be made quite clean for the finish coat to be loaded into the cup. The spraying of this coat is as for the primer—face and nose masks should be worn; extraction fans should be switched on; spray to an even depth of less than 0·001 in.

Drying: This is as for the primer—the oven set to operate at 90°C and, depending on the size of the article, approximately 15 minutes.

Sintering: Sintering is the stage where a physical change takes place and individual particles of PTFE become fused together. This change occurs at about 330°C but the oven should be set to operate between 380°C and 400°C, which will accelerate the sintering process. Again, time will largely be based upon the mass of the object and its thermal conductivity. However, a cast aluminium-alloy frying pan of 200 mm diameter has been found to sinter satisfactorily for 10 minutes at the 380–400°C oven setting.

Cooling: This is the final stage of coating: the temperature encountered at this point is by far the highest met with so far in school work in plastics, so be prepared for it—have tongs ready to hold the article and wear asbestos gloves as a safety precaution. The cooling from what will be approximately 400°C should be gradual, even retarded; it should not be accelerated by draughts or quenching. Rapid cooling will cause the coating and the metal substrate to contract rapidly and breakdown of the coating will result. At all times during coating try to achieve a dust-free atmosphere, for dust and grit will adhere to wet PTFE and will mar the finish.

Safety Notes

1 Read Leaflet F.10 published by Imperial Chemical Industries Ltd, before attempting any work with PTFE in dispersion form.

2 Remember sintering temperatures are high—be prepared for them!

3 As in all spraying work, protect the eyes, nose, mouth and lungs by wearing adequate face masking.

157

5 Machining of Plastics

Introduction

There is today a wide diversity of plastics materials which can and do replace the conventional materials, such as wood and metal. Alongside mild steel, brass and aluminium, plastics such as nylon which can produce an equal if not better performance than their metallic counterpart are in use in school. Such plastics materials add a new dimension to practical work, both artistically and technically, and students may learn the techniques of shaping them with the help of the usual workshop machinery and equipment. Most of the machining techniques are very similar to the ones employed for the shaping of metals and so students should not find any real difficulty in adjusting to the new medium. A change from machining brass to machining aluminium requires an adjustment of speeds, feeds and tool angles, and similar adjustments are made when switching to the machining of plastics materials. It is true to say that it is only when a range of plastics materials is available in the school workshop that their usefulness and comparability with the more conventional materials can be seen.

Materials Available for Machining

Most plastics which can be machined fall into the thermoplastic materials group. The following can be used in a wide range of applications:

Nylon	Cellulose acetate
Polyacetal	Polypropylene
Polycarbonate	PTFE

These materials are usually available in rod, tube, sheet and extruded section form.

1 Nylon

This is a polymer of the polyamides group. Many different types of this material exist, all denoted by a number, e.g. Type 6.6, Type 6.8, Type 6.10, corresponding to the chemical composition. A manufacturer's colour code corresponds to this number so the type of nylon can be recognized quickly by looking for the coloured stripes that are marked on each length of material. All types of nylon are hygroscopic—that is to say they will absorb or lose moisture in air depending

Plate 65   *The range of materials suitable for machining*

on the relative humidity, but material which is stabilized at a relative humidity (R.H.) of 65 per cent can be supplied by the manufacturers. However, if nylon is to be used for a job where accurate dimensions are necessary, this point about moisture absorption must be borne in mind.

Manufacture Details: All rod, sheet, tube and extruded sections in nylon are made from chips of the material by an injection-moulding or extrusion process.

Colour: White or cream are the usual colours, but a range of other colours including black can be obtained; the black pigmented type may contain molybdenum disulphide and is recommended for the manufacture of bearings.

159

General Characteristics: Nylon is a rigid, hard material, very resilient and light in weight. It possesses good wear and frictional properties, together with a high melting point and a high degree of resistance to chemical attack.

Applications

General: Silent running gears; self-lubricating bearings; bushes—motor-car spring bushes; cams; nuts, bolts and washers; wheels, rollers and bearings.

School: All those listed above and such items as mallet heads, handles, bearing mounts.

Machining Details

Lathework: Treat nylon as hard brass—use a small rake angle, good front and side clearance. Employ coolant to avoid heat build-up—heat will cause dimensional instability. High

Plate 66   *Items machined from Nylon 66 stock shapes*
Courtesy of Polypenco Ltd

speeds can be used—up to 1000 ft per minute—but keep cuts shallow to avoid stresses being set up.

Drilling: Drilled holes tend to be undersize due to the material 'giving' somewhat under the pressure of the cut. If necessary, drills can be ground with the point slightly off centre to correct this tendency. Use coolant and clear the swarf frequently from the hole and the drill flutes.

Threading: Again, undersize holes and hence undersize threads can result when nylon is screwed or tapped. Use a larger tapping drill than for metals, and a lubricant to avoid heat build-up and hence an expansion of the material.

Sawing: Use coarse-tooth saws or wood hand-saws.

Safety Notes
In turning and drilling beware of swarf building up and becoming wrapped around the chuck. This can be very dangerous for the operator.

2 Polyacetal

This material is a highly crystalline form of polymerized formaldehyde in a resin form. Its qualities are that it possesses very low moisture absorption and good dimensional stability. Because it absorbs very little moisture, it presents less dimensional variations than nylon. This is the main advantage of polyacetal over nylon, for in practically all other respects the two materials are the same.

Manufacture Details: It is produced as nylon is produced—that is, moulded or extruded from chips of the material—and it is available in rod, sheet, tube and plate forms.

Colours: It is usually a white/cream colour. Black pigmented material is produced as this colour offers resistance to ultra-violet light; special colours are produced to order.

General Characteristics: It is a very stable material and is stiff, strong and quite hard. Like nylon its coefficient of friction is very low and it also possesses good abrasion resistance. Parts fabricated from polyacetal need no lubrication provided they are run in initially with some lubricant.

Applications
General: Gear wheels; shafts and bushes; bearings; cams; instrument parts.

161

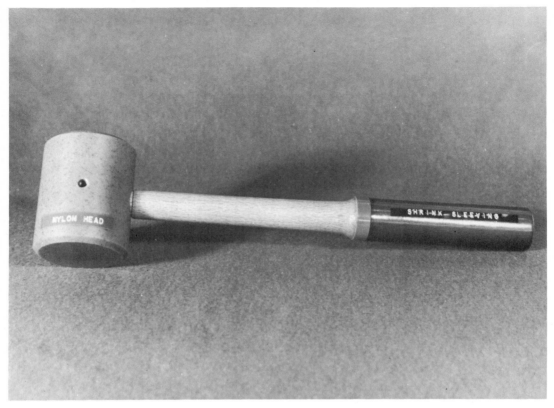

Plate 67　　　　　　　　*A mallet with nylon head machined in school*

School: All the above applications apply to the school workshop. Many other parts normally made out of a metal may be made just as successfully and perhaps more easily in this material.

Machining Details: Polyacetal can be machined just as easily as any of the nylon grades, probably more easily, in fact, for being a harder, stiffer material it tends to 'give' less than nylon and is less inclined to burr over when a cut is being taken. The machining techniques quoted already for nylon can be applied equally successfully to polyacetal.

Safety Notes: As for nylon, beware of any long lengths of turnings becoming wrapped around the moving parts of the machine, thus exposing the operator to a degree of danger.

Plate 68   *Swarf round a lathe chuck*

3 Polycarbonate   This is a thermoplastic material which can be moulded from the powder form or fabricated from the solid state. It is a comparatively new material and it is also produced in a reinforced form called Glass-filled Polycarbonate, the performance of which is held to be superior to the unreinforced material.

Manufacture Details: It has a limited use in fabrication techniques and is really more of a moulding material. It is produced in rod and sheet forms.

Colour: In the natural state it is a very pale yellow or amber colour and could be mistaken for cellulose acetate. It can be pigmented should that be necessary.

163

General Characteristics: It is hard and a material which is stable dimensionally. It possesses a high impact strength and very low moisture-absorption figure—of the order of 0·4 per cent maximum. It keeps its mechanical strength over a wide range of temperature and all these properties commend it for many applications. It is attacked rather easily by some organic solvents, and when large amounts of material are machined away stresses are set up which necessitate the application of a stress-relieving process after the machining of a component.

Applications

General: Because it possesses good dielectric properties, it is used often in electrical components. In film form, it is used in capacitor manufacture. In moulding powder form, it can be used to make bottles and protective headgear. Chosen for its good impact strength values.

School: Rather limited applications—handles, knobs, etc., where temperatures are liable to vary from the normal.

Machining Details: Much of the information given for nylon applies to this material, but because of the stress inducement that can take place, all corners on a component need to be radiused, all the cuts must be light, and in any threading work single-point tools rather than conventional taps and dies should be used. Avoid coolants that contain any solvent liable to attack the material. Cracking and crazing of the surface both denote attacks by solvents.

4 Cellulose Acetate

This material has been developed from cellulose nitrate, the highly inflammable material. The acetate material is 'non-flam' and therefore safe to use. Both cellulose nitrate and cellulose acetate are derived from the naturally occurring polymeric material cellulose which is found in wood pulp and cotton linters.

Manufacture Details: It is produced in very thin films, sheets and rods. It can be extruded and this is the method used to produce sheets, rods and sections; thin foils and films can be cast from a solution.

Colour: Clear transparent and a range of colours in sheet and film forms. The rod is commonly available in a yellow/amber shade; produced in a variety of fluted shapes.

General Characteristics: It has a high impact strength which combines with a good low-temperature performance. It is a hard rigid material, resilient and light in weight. It has quite a low softening-point temperature and this allows it to be friction welded easily. Its water absorption is quite high and this affects its electrical insulation properties.

Applications

General: The rod is used extensively these days for tool handles—chisel, screwdriver—because of its good impact strength, resilience and lightness. In sheet and film form it is used much in the fabrication of guarding for machinery, masks, visors and goggles.

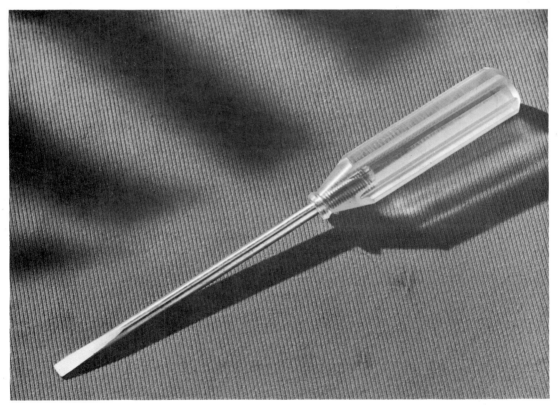

Plate 69

A screwdriver with cellulose acetate handle machined in school

School: A wide range of tool handles can be made—the rod is available in cylindrical and fluted forms and can be turned into a variety of designs.

Machining Details: The material machines very well and the techniques for nylon and the other materials mentioned so far apply. However, it does possess a low softening temperature, so heat build-up must be avoided in any machining. Keep tool edges keen and well honed and the finish will be all the better.

Safety Notes: These are as for the machining of the other materials, for long pieces of machinings can build up and become wrapped around the moving parts of the machine. Do not friction weld on a fast speed as the material can go out of control.

5 Polypropylene

This is a plastic of the polyolefin group and is used much today where polyethylene was formerly used. It has properties very akin to nylon, and because of its resistance to certain acids it can be used where nylon would break down. It combines rigidity, even at the point of boiling water, good electrical insulating properties and safety when in contact with foodstuffs. It is certainly a versatile material!

Manufacture Details: It is normally available in sheet and rod form—the sheet is extruded and rolled, the rod simply extruded. The sheet can be produced embossed with patterns by a further stage of rolling and it will retain such embossing during heating for subsequent shaping.

Colour: Polypropylene in its natural unpigmented form is pale pink/white in colour but it can be pigmented in a range of colours.

General Characteristics: As has been already stated, polypropylene is a versatile material possessing many qualities —rigidity, high impact strength and good mechanical and electrical properties. Shapings in this material will withstand boiling water and such acid solutions as hydrochloric and sulphuric, whereas such materials as nylon will be attacked. The surface being chemically inert, it can only be welded with hot-air gun and welding rod; no adhesive will take to it nor will paint. This material has the unique ability of being able to be

166

formed into hinges which can be flexed many times before showing signs of failure or fracture.

Applications
General: Boilable hospital and laboratory ware; radio and television cabinets; high heels for ladies' shoes.

School: Hinges for boxes and doors; fabrication of trays; handles and mallet heads.

Machining Details: All the techniques used in metalworking apply—sawing, drilling, milling, turning, using conventional tools and maintaining sharp, keen edges. Cutting lubricants should not be needed. Much of the machining procedure for nylon applies for this material.

6 PTFE (Poly-tetrafluoro-ethylene)

This is the most fascinating of all the plastics materials as it possesses so many attributes to commend its use. It has the lowest coefficient of friction of any solid material. It has the greatest resistance of any plastics to chemical attack. It has excellent electrical insulation properties. Its only shortcoming is that under load high plastic deformation can take place. Its well-known trade names are 'Fluon' and 'Teflon' and it is widely used as a dispersion coating on saucepans, frying pans and cooking utensils in general (see Section V, 4).

Manufacture Details: Unlike other thermoplastic materials, its manufacture is not by the conventional injection-moulding or extruding processes. It is produced by a sintering process which has been demonstrated earlier under 'Use of PTFE'. The material, in powder form, is compacted into a mould form when cold and then heated to its sintering temperature, approximately 330°C. It is then cooled slowly to obtain stability and even crystallinity in the finished form. It is generally available in rod, sheet, tube, sleeving and tape forms.

General Characteristics: PTFE, although belonging to the vinyl group of plastics, is similar to polyethylene in that it has a 'waxy' surface, is taste-free and is odourless. PTFE is also 'non-flam'. The common colour is white with dispersions made in grey and black pigments. It can be freely machined as other thermoplastics, but cannot be shaped under heat and pressure.

167

Applications

General: Seals and gaskets; pressure-valve parts; insulators for high-grade electrical work; non-stick surface rollers; coating metals to be in contact with foodstuffs and liquids.

School: Bearings; friction experiments; coating on utensils.

Machining Details: Similar methods apply as for nylon, although as PTFE exhibits deformation under load, the material must be supported adequately during machining and tool edges must be kept sharp with polished surfaces.

Section VI

1 Glossary of Terms
2 Identification of Plastics by Simple Tests
3 Suppliers of Equipment and Materials
4 Addresses for Literature and General Bibliography
5 Proprietary Names and Manufacturers

1 Glossary of Terms

ABS : An abbreviated form for acrylonitrile-butadiene-styrene. Available as a moulding material for extrusion and injection moulding and in sheet for vacuum forming and blow moulding.

Accelerator : A substance with a catalysed resin to increase the efficiency of the 'catalyst'. It is sometimes known as an activator or a promoter.

Acrylic : A synthetic resin prepared from acrylic ester monomers. Common trade names are 'Perspex', 'Lucite', 'Plexiglas' or 'Oroglas'.

Activator : See *Accelerator.*

Bag Moulding : Shaping by the application of pressure during bonding or moulding, in which a flexible cover exerts pressure on the material to be moulded, through the introduction of air pressure or drawing off air to create a vacuum.

Blow moulding : A forming operation, usually for bottle production, that forces air into a tube of heated plastics situated in a split mould.

Calendering : A process where a warm doughy mass of plastics material is passed between a series of rollers and emerges as flat film or sheet.

Casting : The formation of an object by pouring a fluid resin into a mould.

'Catalyst' : A chemical compound usually added in small quantity to a monomer to speed up polymerization.

Celluloid : The first modern plastics material. A mixture of camphor and cellulose nitrate.

Cellulose : A naturally occurring polymer found in wood pulp and cotton linters. Some derivatives are cellulose nitrate, cellulose acetate and cellulose acetate butyrate.

Cold moulding : A process using high pressure to cold form a filled and uncured thermosetting plastics, subsequently cured by heating.

Compression moulding: The forcing of a raw plastics, usually of the thermosetting variety, into the required shape under the influence of heat and pressure.

Cure: The change of physical properties by chemical reaction. Usually achieved by the addition of heat and/or a 'catalyst' and with or without pressure. Also known as **Set**.

Dielectric: Any material that will resist the passage of an electrical current; such a material is a good insulator and most plastics are good insulators.

Dip coating: The application of a plastics coating by dipping a heated article into a tank of fluidized thermoplastics powder and then fusing the powder adhering to the surface. A liquid or paste can also be used, employing a similar dip and fusing technique. This process is also known as **Plastics coating**.

Dispersion: Finely divided particles of a material in suspension in another substance.

Dowel pins: Steel pins in one part of a mould, holding it in position with another part.

Draught: The degree of taper of a side wall or the angle of clearance in a mould to facilitate the removal of parts from the mould.

Expanded plastics: Sometimes called foamed plastics, plastics foams or cellular plastics. Expanded by gas introduced chemically or mechanically which leaves a cell-like structure in the material.

Extrusion: An operation in which powdered or granular material is fed through a hopper, heated and carried forward by a revolving screw, eventually being forced through a shaping orifice. Similar to the mincing of meat in a mincer.

Fillers: Inert substances added to plastics materials to make them less costly and to improve physical properties such as hardness and stiffness.

Flash: Extra plastics material attached to a moulding along the parting line.

Flow: The fluidity of plastics material usually during a moulding process.

171

Foaming agents: Chemicals added to plastics and rubbers that generate inert gases when heated, causing the material to assume a cellular form.

Gate: A small restricted channel in an injection mould linking the sprue and the mould cavity.

Gel: A semi-solid system consisting of a network of solid aggregates in which a liquid is held.

Gel coat: A thin outer layer of resin, sometimes containing pigment, which is applied to a mould prior to reinforcing with glass fibre.

Glass fibre and **Glass cloth:** Reinforcing materials used with polyester resins. Fibres of glass approximately 0·00025 in diameter are laid in mat form, held in place with adhesive or are woven into a variety of textured cloths.

Granules: Small cylindrical-shaped pieces of plastics material made by the extrusion process and used in moulding or sintering processes.

Hardener: A substance added to a polymer or synthetic resin to promote curing. See *'Catalyst'.*

Injection moulding: A process that involves the feeding of granules of plastics material into a heated cylinder, where a ram or screw forces the melted material through a nozzle where it is injected into a mould.

Laminate: Superimposed layers of resin-impregnated material which are bonded together by means of heat and pressure to form a single piece.

Lay-up: The process of placing reinforcing material into position in a mould.

Light piping: The ability of optical glass or polished acrylic sheet to pass light from one end to another with little loss, even around bends. Also known as **Edge lighting**.

Melamine: The full name is melamine-formaldehyde resin or MF resin. A thermosetting plastics used in laminate work and in the production of table ware.

Methyl methacrylate: Chemical name for one acrylic monomer.

Monomer: A relatively simple compound which can react to form a polymer.

Mould release agent: A lubricant used to coat the mould surface to prevent the plastics material sticking to it.

Nylon: The generic term for all synthetic polyamides.

Parting agent: See *Mould release agent.*

Parting line: The mark around a moulding where two parts of a mould meet.

Phenolic: A thermosetting resin produced by condensation of the phenol-formaldehyde type.

Plastics: A term given to a synthetic material based on an organic polymer which can be permanently formed or deformed under external stress or pressure, usually accelerated by the use of heat.

Plastics coating: See *Dip coating.*

Plastic memory: The quality of some thermoplastics materials to return to their original form after re-heating. Also known as *Elastic memory.*

Plasticizer: An additive which makes the plastics material pliable, softer and easier to mould into shape.

Plastisol: A suspension of finely divided resin in a plasticizer. The resin usually dissolves at elevated temperatures and a homogeneous plastic mass results when the mixture cools.

Polyacetal: A hard polyformaldehyde material possessing low coefficient of friction and good machining properties. It is replacing metal in many instances.

Polycarbonate: One of the polyester group having high dimensional stability and very good impact strength.

Polyester: A thermosetting resin of the unsaturated type used in conjunction with styrene monomer as the binding resin in glass-fibre reinforced plastics.

Polyethylene: A thermoplastics material generally divided into two groups—low density and high density. The former is manufactured under elevated temperatures and high pressures and the latter at low or medium temperatures and pressures.

Polymer: The combined or joined smaller molecules that have formed larger molecules.

Polymerization: The act of combining two or more molecules into a single larger molecule.

Polypropylene: A member of the polyolefin group, as is polythene, but with a better heat and chemical resistance than polythene. It is used extensively in injection moulding and in sheet and film production.

Polystyrene: A water-white thermoplastics of the vinyl group. It needs to be toughened to reduce brittleness and is widely used in injection moulding. In the expanded form, it has broad applications in the insulation and packaging fields.

Polyurethane: A thermosetting material used a great deal in foam form which, when hardened off, can be cut and shaped. In the foam state, it can be used to insulate, 'pot' electrical parts and provide buoyancy in boats.

Polyvinyl acetate: In the vinyl group, similar to PVC but its main use is in dispersion form as an adhesive or a coating agent. Known by the initials PVAC.

Polyvinyl chloride: Produced in rigid, plasticized and dispersion forms and has a variety of applications—rainwater goods, bottle production and coating fabrics. Known by the initials PVC.

Pot life: See *Shelf life.*

PTFE: Polytetrafluoroethylene is in the vinyl group. It is an expensive material offering an extremely low coefficient of friction, excellent insulating properties and chemical inertness. One of its many uses is in the coating of non-stick surfaces on kitchen ware.

Shelf life: The length of time a plastics material will remain a liquid before curing—usually during storage.

Sintering: The process of holding powders or granules at just below their melting point; the particles are then fused or stuck together but not melted.

Slush moulding or **casting:** A method of casting in which a liquid, usually PVC paste, is poured into a heated mould where a viscous skin forms. Excess paste is poured off and the

174

skin cured by additional heating. The moulding is then removed.

Solvent: Any substance that dissolves other substances.

Thermoplastics: A material that softens when heated, and may be shaped to a desired form which is retained upon cooling to the solid state. This process can be repeated a number of times.

Thermosetting plastics: A material which, when heat and pressure are applied to it, hardens or cures to a solid mass. The process is irreversible and the material will not soften under further heat and pressure.

Transfer moulding: Similar to compression moulding, but the material, which is usually of the thermosetting variety, is pre-heated in an antechamber until it flows and is then injected under pressure into a heated mould. By this technique the mould is subjected to less wear.

Urea-formaldehyde: A thermosetting material known as UF resin. Used in a filled form, it is satisfactory for moulding and has applications in adhesives for wood.

Vacuum forming: A method of forming thermoplastics sheet. The sheet is clamped in a frame, heated and drawn down onto a mould form by vacuum pressure.

Welding: The joining of thermoplastics materials by one of several softening process—hot-gas torch, heated rollers or strips, friction or high-frequency heating.

2 Identification of Plastics by Simple Tests

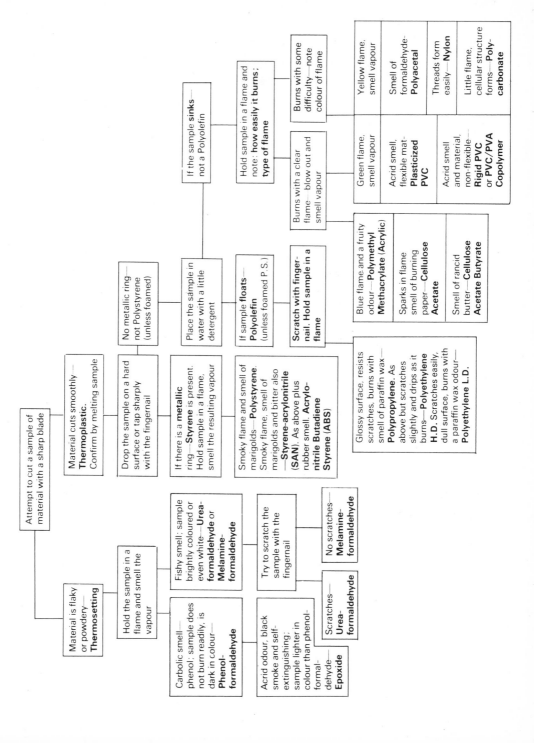

3 Suppliers of Equipment and Materials

Tile Making

Low-density Polyethylene Powder
Griffin & George Ltd

Trylon Ltd, WOLLASTON, Northants., NN9 7QJ

Adhesives
Araldite AZ 107 and Hardener HZ 107
CIBA (A.R.L.) Ltd, DUXFORD, Cambs.
Evostick Impact Adhesive
Evode Limited, STAFFORD

Ovens
Griffin & George Ltd

De-greasing Solution—1 : 1 : 1-Trichloroethane
Griffin & George Ltd

Resin Moulding

Resin, Accelerator, Catalyst, Pigments
Griffin & George Ltd

Trylon Ltd, WOLLASTON, Northants., NN9 7QJ

Powder Paint
*Margros Ltd, Monument House, Monument Way West,
WOKING, Surrey*

Paraffin Wax, Beakers, Stirring Rods, Acetone
Griffin & George Ltd

Trylon Ltd, WOLLASTON, Northants., NN9 7QJ

Microcrystalline Wax
*James B. Berry's Sons Company Ltd, 7 Woodcote Road,
WALLINGTON, Surrey*

Glass Matting and Cloth
Griffin & George Ltd

Trylon Ltd, WOLLASTON, Northants., NN9 7QJ

Turner Brothers Asbestos Co. Ltd, ROCHDALE, Lancs.

Acrylic Materials	'Perspex'

ICI 'Tensol' Cements for 'Perspex'
G. H. Bloore Ltd, 480 Honeypot Lane, STANMORE, Middlesex
'Oroglas' Material
Griffin & George Ltd

Toggle Clamps for Machining Sheet—'De-Sta-Co'
Insley Industrial Limited, Eastern Road, BRACKNELL, Berks.

'Inconel' Heating Elements
AEI Heating Ltd, Redring Works, PETERBOROUGH, Northants.

or *your local Electricity Board (Cooker Elements)*

Ovens
Griffin & George Ltd

Dip or Plastics Coating

Low-density and High-density Polyethylene, PVC Powder, Rilsan (Nylon) Powder, Epoxy Resin Powder, Cellulose Acetate Butyrate
Griffin & George Ltd

Low-density Polyethylene
Trylon Ltd, WOLLASTON, Northants., NN9 7QJ

Fluidized Bed Equipment
Griffin & George Ltd

Pyrolith Tiles for Fluidized Bed Equipment
Doulton Industrial Products, Aerox Filtration Div., Filleybrooks, STONE, Staffs.

Air Compressors (B.E.N. Handispray Unit)
Broom & Wade Ltd, P.O. Box 7, HIGH WYCOMBE, Bucks.

De-greasing solution—1:1:1-Trichloroethane
Griffin & George Ltd

Centrifugal Casting

Large-diameter Copper Tube
Stuart Turner Ltd, HENLEY-ON-THAMES, Oxon.

Polyester Resin, Accelerator, Catalyst, Pigments
Griffin & George Ltd

178

Trylon Ltd, WOLLASTON, Northants., NN9 7QJ

Powder Paint
*Margros Ltd, Monument House, Monument Way West,
WOKING, Surrey*

Beakers, Stirring Rods, Acetone
Griffin & George Ltd

Trylon Ltd, WOLLASTON, Northants., NN9 7QJ

Compression Moulding

Small Compression Moulding Machines (Sold for Metallurgical Mounting)
*Buehler Ltd; 2120 Greenwood Street, P.O. Box 830,
Evanston, Illinois, U.S.A.*

Compression Moulding Machines
Griffin & George Ltd

Heater Bands
*A.E.I. Heating Ltd, Redring Works, PETERBOROUGH,
Northants.*

Hedin Heat Ltd, Raven Road, LONDON, E18

Bakelite Moulding Material
*Metallurgical Services Ltd, Reliant Works, BETCHWORTH,
Surrey*

Phenolic, Melamine-formaldehyde and Urea-formaldehyde Moulding Materials
Griffin & George Ltd

Blow Moulding

Acrylic, Cellulose Acetate, PVC
Griffin & George Ltd

Extrusion

Machines and Materials
Griffin & George Ltd

Fabrication from Expanded Polystyrene

Hot-wire Machines
Almik Displays Ltd, 10 Wendell Road, LONDON, W12 9RT

Expanded Polystyrene
Almik Displays Ltd

179

Trylon Ltd, WOLLASTON, Northants., NN9 7QJ

Polyurethane Foam
Trylon Ltd, WOLLASTON, Northants., NN9 7QJ

Adhesives for Expanded Polystyrene: Tretobond 740,
Tretobond 282, Tretobond 375

Tretobond Ltd, Tretol House, The Hyde, LONDON, N.W.9

Ni-Chrome Wire for Hot-wire Machines
Griffin & George Ltd

Electrical Fittings for Hot-wire Machines
Radiospares, P.O. Box 427, 13–17 Epworth Street, LONDON, E.C.2

Beads for making Expanded Polystyrene—'Styrocell'
Griffin & George Ltd

Vacuum Forming

Vacuum-forming Machines
Griffin & George Ltd

Trylon Ltd, WOLLASTON, Northants., NN9 7QJ

Shoe and Allied Trades Research Association, Satra House, Rockingham Road, KETTERING, Northants.

General Sheet Requirements
Trylon Ltd, WOLLASTON, Northants., NN9 7QJ

Pumps, Pipes and Pipe Fittings for Vacuum-forming Machines
Enots Ltd, Aston Brook Street, BIRMINGHAM 6

Electrical Elements for Vacuum-forming Machines—'Inconel'
Sheathed Elements
A.E.I. Heating Ltd, Redring Works, PETERBOROUGH, Northants.

or *your local Electricity Board (Cooker Elements)*

Acrylic, Cellulose Acetate, PVC Sheet
Griffin & George Ltd

Injection Moulding

Injection Moulding Machines
The Small Power Machine Co. Ltd, Bath Road Industrial Estate, CHIPPENHAM, Wilts.

Temperature Gauges, Dial Type for Machines
*The British Rototherm Co. Ltd, Merton Abbey, LONDON,
S.W.19*

Face Masks, Asbestos Gloves
Griffin & George Ltd

Moulding Materials: Acrylic, ABS, Nylon, Polypropylene,
Polystyrene, Polyethylene, PVC
Griffin & George Ltd

*The Small Power Machine Co Ltd, Bath Road Industrial Estate,
Chippenham, Wilts., SN14 0BR*

**Welding of
Plastics**

Roller Plastics Sealing Tool and 45-W Elements
*The Acru Electric Tool Mfg Co. Ltd., Acru Works,
Demmings Road, CHEADLE, Ches.*

Hot-gas Welding Equipment
Bielomatik London Ltd, 7 Cotswold Street, LONDON, S.E.27

Polyethylene Sheet and Lay-flat Tubing
*Transatlantic Plastics Ltd, 45 Victoria Road, SURBITON,
Surrey*

Sheet Materials and Welding Rods
G. H. Bloore Ltd, 480 Honeypot Lane, STANMORE, Middlesex

Barrier Film—ICI 'Melinex' Polyester Film
Griffin & George Ltd

Trylon Ltd, WOLLASTON, Northants., NN9 7QJ

Heat-sealing Equipment
*Theco Electrical Appliances Ltd, 4–10 Wakefield Road, South
Tottenham, LONDON, N.15*

Use of PTFE

PTFE Dispersions
Griffin & George Ltd

Spray Guns
*The DeVilbiss Co. Ltd, Ringwood Road, BOURNEMOUTH,
Hants.*

Breathing Masks
Griffin & George Ltd

Ovens (Pottery Kilns)—Sintering
Wengers Ltd, STOKE-ON-TRENT, Staffs.

Ovens (for Low-temperature Work)
Griffin & George Ltd

Machining Plastics

Materials—Nylon, Polycarbonate, Polyacetal, Polyethylene, Polypropylene, PTFE
Nylonic Engineering Ltd, Woodcock Hill, RICKMANSWORTH, Herts.

G. H. Bloore Ltd, 480 Honeypot Lane, STANMORE, Middlesex

Griffin & George Ltd

Cellulose Acetate Rod
Harold Moore & Son Ltd, Bailey Works, Bailey Street, SHEFFIELD, 1S1 3BR

Griffin & George Ltd

4 Addresses for Literature and General Bibliography

Information

1 Academic Liaison Officer, Imperial Chemical Industries Ltd,
Plastics Division, P.O. Box 6,
Bessemer Road, WELWYN GARDEN CITY, Herts.

2 BP Educational Service,
The British Petroleum Company Ltd,
Britannic House, Moor Lane,
LONDON, EC2Y 7BU

3 Shell Chemicals U.K. Ltd,
Shell Centre,
LONDON, S.E.1

4 British Industrial Plastics Ltd,
P.O. Box 6, Popes Lane,
OLDBURY, Warley, Worcs.

5 Bakelite Xylonite Ltd,
Enford House, 139 Marylebone Road,
LONDON, NW1 5QE

6 The Plastics and Rubber Institute,
11 Hobart Place,
LONDON, SW1W 0HL

7 Trylon Ltd,
Thrift Street, Wollaston,
WELLINGBOROUGH, Northants., NN9 7QJ

8 The British Plastics Federation,
47 Piccadilly,
LONDON, W1V 0DN

9 CIBA (A.R.L.) Ltd, Plastics Division,
DUXFORD, Cambs., CB2 4QA

10 Fibreglass Limited, ST HELENS, Lancs.

Books

Organic Chemistry through Experiment by D. J. Waddington and H. S. Finlay (Mills & Boon)

Experiments in Polymer Chemistry by H. S. Finlay, B.Sc. (Shell International Petroleum Co. Ltd)

Plastics as an Art Form by Thelma R. Newman (Sir Isaac Pitman & Sons)

Plastics for Engineers by G. R. Palin, B.Sc., Ph.D. (Pergamon Press)

Plastics and You by R. Lushington (Pan Books)

183

Practical Work with Modern Materials by John Robinson (Edward Arnold Ltd)

Design Engineering Handbook—Plastics (Morgan-Grampian Ltd)

Plastics in the Modern World by Couzens and Yarsley (Pelican Books)

The Elements of Injection Moulding of Thermoplastics (Learning Systems Ltd)

Introduction to Plastics by Briston and Gosselin (Newnes Books)

Your Guide to Plastics by J. Gordon Cook (Merrow Publishing Co. Ltd)

A Journalist's Guide to Plastics by R. C. Penfold (The British Plastics Federation)

A First Look at Plastics by O. J. Walker (The Plastics Institute)

Plastics, Rubbers and Fibres by L. W. Chubb (Pan Books)

The Identification of Plastics and Rubbers by K. J. Saunders (Chapman & Hall)

Experimental Plastics by Redfarn and Bedford (Iliffe Ltd)

A Guide to Plastics by C. A. Redfarn (Iliffe Ltd)

Working of Plastics Manual—Organization for Economic Co-operation and Development (H.M.S.O.)

The British Plastics Year Book—A Directory published annually (Iliffe Industrial Pub. Ltd)

The Story of B.I.P. by Dingley (British Industrial Plastics Ltd)

Plastics by Dubois and John (Reinhold)

Landmarks of the Plastics Industry (Imperial Chemical Industries Ltd, Plastics Division)

B.S. 3502—*Standard Names of Plastics* (British Standards Institution)

B.S.2782—*Methods of Testing Plastics* (British Standards Institution)

The First Century of Plastics by M. Kaufman (The Plastics Institute)

Plastics Materials by J. A. Brydson (Iliffe & Van Nostrand)

The Identification of Plastics by Simple Tests—Technical Service Note G104 (Imperial Chemical Industries)

184

Glass Fibre for Schools—J. Tiranti (Scopas Handbook)

Setting in Clear Plastic by K. Zechlin (Mills & Boon)

Glass, Resin and Metal Construction by Peter Tysoe (Mills & Boon)

New Materials in Sculpture by H. M. Percy (Tiranti)

Plastics by M. Guest—Young Scientist Series (Weidenfield & Nicholson)

The Story of Plastics (Ladybird Books)

Creative Technology, Books 1—4, by Peter Clarke (Allman)

Design with Plastics—Project Technology Handbook (Schools Council)

Children and Plastics Stages 1 and 2 and *Background—Science 5–13 Project* (Macdonald)

Modern Materials for Workshop Projects by Birden and Hilsum (Hutchinson Educational)

5 Proprietary Names and Manufacturers

Resins and Moulding Materials

Acetal	Delrin	E I du Pont de Nemours Co. Inc. (Du Pont Company (U.K.) Ltd)
	Kematal	Imperial Chemical Industries Ltd
Acrylics	Diakon	Imperial Chemical Industries Ltd
	Oroglas	Lennig Chemicals Ltd
Acrylonitrile-butadiene-styrene (ABS)	Cycolac	Marbon Chemical Division of Borg-Warner Ltd
	Kralastic	Uniroyal Ltd
	Lustran	Monsanto Chemicals
	Sternite	Sterling Moulding Materials Ltd
Alkyds	Bakelite	Bakelite Xylonite Ltd
	Epok	BP Chemicals International Ltd

Aminos (see Urea-formaldehyde and Melamine-formaldehyde)

Cellulose Acetate	Dexel	British Celanese Ltd
	Tenite	Eastman Chemical International
Cellulose Acetate Butyrate	Tenite	Eastman Chemical International
Epoxide	Araldite	CIBA (A.R.L.) Ltd
	Bakelite	Bakelite Xylonite
	Epikote	Shell Chemicals U.K. Ltd
	Epophen	Borden Chemical Co. (U.K.) Ltd
	Telcoset (powders)	Telcon Plastics Ltd
Ethylene-vinyl-acetate	Alkathene	Imperial Chemical Industries Ltd
	Elvax	E I du Pont de Nemours Co. Inc. (Du Pont Company (U.K.) Ltd)
	Montothene	Monsanto Chemicals Ltd
Melamine-formaldehyde	Melmex	British Industrial Plastics Chemicals Ltd
Phenolics	Bakelite	Bakelite Xylonite Ltd
	Epok	British Petroleum Chemicals (U.K.) Ltd
	Nestorite	James Ferguson & Sons Ltd
	Sternite	Sterling Moulding Materials Ltd

Polyamide (Nylon)	Maranyl	Imperial Chemical Industries Ltd
	Rilsan	Aquitaine Organico
	Zytel	E I du Pont de Nemours Co. Inc.
		(Du Pont Company (U.K.) Ltd
Polycarbonate	Makrolon	Farbenfabriken Bayer AG
		(J. M. Steel & Co. Ltd)
Polyester	Bakelite	Bakelite Xylonite Ltd
	Beetle	British Industrial Plastics Chemicals Ltd
	Cellobond	BP Chemicals International Ltd
	Crystic	Scott Bader & Co. Ltd
Poly(4-methylpentene-1)	TPX	Imperial Chemical Industries Ltd
Polypropylene	Carlona P	Shell Chemicals U.K. Ltd
	Propathene	Imperial Chemical Industries Ltd
Polystyrene	Carine	Shell Chemicals U.K. Ltd
	Lustrex	Monsanto Chemicals Ltd
	Sternite	Sterling Moulding Materials Ltd
	Styron	Dow Chemical Co. (U.K.) Ltd
	BP Polystyrene	BP Chemicals International Ltd
Polystyrene, Expandable	Montopore	Monsanto Chemicals Ltd
	Styrocell	Shell Chemicals U.K. Ltd
Polythene	Alkathene	Imperial Chemical Industries Ltd
	Bakelite	Bakelite Xylonite Ltd
	Carlona	Shell Chemicals U.K. Ltd
	Rigidex	BP Chemicals International Ltd
	Telcothene	Telcon Plastics Ltd
	(powder)	Monsanto Chemicals Ltd
Polytetrafluoroethylene	Fluon	Imperial Chemical Industries Ltd
	Teflon	E I du Pont de Nemours Co. Inc.
		(Du Pont Company (U.K.) Ltd)
Polyvinyl Chloride	Breon	BP Chemicals International Ltd
	Carina	Shell Chemicals U.K. Ltd
	Corvic	Imperial Chemicals Industries Ltd
	Telcovin	Telcon Plastics Ltd
	(powder)	
	Vinatex	Vinatex Ltd
	Vybak	Bakelite Xylonite Ltd
Styrene Acrylonitrile	Lustran	Monsanto Chemicals Ltd
	Tyril	Dow Chemical Co. (U.K.) Ltd

Urea-formaldehyde	Beetle	British Industrial Plastics Chemicals Ltd
	Nestorite	James Ferguson & Sons Ltd

Sheet and Film

Acrylics	Oroglas	Lennig Chemicals Ltd
	Perspex	Imperial Chemical Industries Ltd
ABS	Iridon	Commercial Plastics Ltd
	Vulkine 'A'	Imperial Chemical Industries Ltd
Cellulose (Regenerated)	Cellophane	British Cellophane Ltd
Cellulose Acetate	Bexoid, Bexfilm	Bakelite Xylonite Ltd
	Celastoid, Cellastine, Clarifoil,	
	Claritex	British Celanese Ltd
	Iridon	Commercial Plastics Ltd
	Acelon	M & B Plastics Ltd
Cellulose Acetate Butyrate	Iridon	Commercial Plastics Ltd
	Cabulite	M & B Plastics Ltd
Cellulose Nitrate	Xylonite	Bakelite Xylonite Ltd
Polyethylene Terephthalate	Melinex	Imperial Chemical Industries Ltd
	Mylar	E I du Pont de Nemours Co. Inc. (Du Pont Company (U.K.) Ltd)
Polypropylene	Bexphane 'P'	Bakelite Xylonite Ltd
	Propafilm	Imperial Chemical Industries Ltd
	Vitralene	Stanley Smith & Company
	Vulkide 'B'	Imperial Chemical Industries Ltd
Polystyrene	Bextrene	Bakelite Xylonite Ltd
	Celatron	British Celanese Ltd
Polystyrene, Expanded	Poron	Poron Insulation Ltd
	Vencel	Benesta
	Warmafoam	Ross Warmafoam Ltd
Polythene	Bexthene	Bakelite Xylonite Ltd
	Marleythene	The Marley Tile Co. Ltd
	Telcothene	Telcon Plastics Ltd
	Vitrathene	Stanley Smith & Company
	Visqueen	British Visqueen Ltd

188

PVC (Unplasticized)	Cobex	Bakelite Xylonite Ltd
	Craylon	Commercial Plastics Ltd
	Darvic	Imperial Chemical Industries Ltd
	Telcovin	Telcon Plastics Ltd
	Vitrone	Stanley Smith & Company
PVC (Plasticized)	Fablon	Commercial Plastics Ltd
	Con-tact	Storey Brothers & Co. Ltd
	Velbex	Bakelite Xylonite Ltd